THE INGENUITY OF MAN

The Ingenuity of Man

by

ESTELLE H. RIES

ILLUSTRATED

GREENWOOD PRESS, PUBLISHERS
WESTPORT, CONNECTICUT

Library of Congress Cataloging in Publication Data

Ries, Estelle H
The ingenuity of man.

Reprint of the ed. published by Exposition Press, New York; originally published in 1930 under title: Mother wit.
Includes index.
1. Civilization--History. 2. Inventions--History.
I. Title.
[CB59.R53 1975] 909 75-8726
ISBN 0-8371-8041-4

SECOND, ENLARGED EDITION

This edition originally published in 1962 by Exposition Press, New York

Reprinted with the permission of Exposition Press, Inc.

Reprinted in 1975 by Greenwood Press,
a division of Williamhouse-Regency Inc.

Library of Congress Catalog Card Number 75-8726

ISBN 0-8371-8041-4

Printed in the United States of America

To the Memory of My Father

ELIAS E. RIES

Fellow, American Institute for Advancement of Science

A GREAT INVENTOR AND A GREAT SOUL

WHOSE MANIFESTATIONS OF INGENUITY GAVE THE IMPETUS

FOR THIS BOOK

Contents

LIST OF ILLUSTRATIONS

THE INGENUITY OF MAN

The Metropolitan Museum of Art, New York

Adam and Eve, an engraving by Albrecht Dürer (1504), probably the most famous engraving of nude figures ever made. Adam names the animals (see page 107).

I

Mother Wit

It is said that the lobster when washed high and dry, makes no effort to get back into the water but waits for the sea to come to him. If it does, well and good; if it does not, he simply dies. This is a total absence of ingenuity. Lobsters have all they need to adjust themselves to such a condition except the ingenuity to use their powers. One of the advantages of being human is that one does not need to wait for the sea to come to him, but, with reasonable ambition and character, may go after what he wants and needs.

Ingenuity, or mother wit, as it is often called, has come to mean the quality or power of ready invention: quickness or acuteness in producing new ideas or relationships. It is the phase of our creative power that is shown in devising expedients, overcoming difficulties, inventing appliances, and adapting means to ends. Genius comes as a flash and is then developed; it gives itself rather to exalted intellectual power in original pure creation for its own sake. But ingenuity is a more deliberate thing. It can be cultivated.

To man, an obstacle has always been a challenge. We do not need to accept obstacles, for by taking thought we may overcome them. Someone asked an old man what he did with his spare time. "I jest sets and thinks, and sometimes I only sets," he replied. We can imagine the difference between the thinking of a man who daydreams in this way, and that of one who really thinks to solve a problem. What nature had denied the primitive in physical powers, he supplied to himself through the exercise of such of his thinking as he could make dynamic and constructive. Seeing the claws of animals, he took the idea of the rake and the hoe. From the woodpecker he visualized the gimlet, from the beaver the trowel, from the rolling log a wheel.

The story behind our commonplaces is a tale of many such adventures in ingenuity, and by knowing them we feel ourselves more a part of the general motion toward progress. New devices and methods come to the fore so rapidly in this inventive age that we tend to believe that everything ingenious arrives in its full perfection almost without parentage or ancestors. But if we survey the heritage of ingenuity, we have the past to build on. We do not in each effort have to dig for new foundations; we can begin with all there is. By following these stories of man's ingenuity we see by what apparent trifles, what tricks of observation or analysis, good things are evolved, and thereby our own mother wit is best stimulated. There is a certain extension of one's self by virtue of such creativeness. By analyzing and improving the conditions, we are no longer an outsider, the tool of circumstances, but feel ourselves a part of them—or, better yet, their master. This does not occur of itself by our being merely passive and submissive. Only by contributing the proper effort is the change wrought.

Change is the friend of progress. Today a chemist turns a reeking, sputtering test tube in the blue flame of a Bunsen burner; tomorrow, in a world-wide industry, a worthless by-product will become a profitable staple. This morning an Edison sits under a hooded light in his workshop, about him a litter of crude sketches, a parade of penciled figures; and presently, in a hundred thousand factories, one man will do the work of fifty. So comes progress. But even genius depends on the data within reach, the knowledge of what others have done. Aristotle, great thinker of the classical world, could not have invented Edison's inventions, because science had not provided the source material. Change, however, is not always progress. Experiments are sometimes costly, but changelessness is sure to be backward-looking.

Movement is the essence of everything. Even a Utopia which is stationary at the highest possible level is by that very virtue a deadly thing. With nothing more to strive for, with no reason for ambition or aspiration or passion or emotion, with nothing to do because everything has been done, there could be neither mental comfort nor spiritual peace. Life would grow unendurable. But there is no immediate danger!

Ingenuity is regrettably often retrogressive, as in the case of inventiveness directed toward war. Yet many examples of ingenuity that are not of themselves ideal, as in the story of sanitation, may perhaps be good in relation to what went before. Perhaps a few centuries hence, who shall say but that our modern methods shall have been equally excelled as we have excelled the past? Progress, we see, is not always forward. Like the waves of a rising tide, each wave comes a little farther up the beach. But between it and the next there is an interval where the old mark is wiped out and the new not yet fixed. At those times mother wit seems to be storing up energy for future use.

But in some way ingenuity represents man's ambitions, and in a finer sense his hopes and aspirations. Tagore, the great Hindu philosopher, very beautifully sounds the note of our native optimism in a passage from *Sadhana* in which he says: "Man does not really believe in evil, just as he cannot believe that violin strings have been purposely made to create the exquisite tortures of discordant notes, though it can be mathematically proved that the probability of discord is far greater than that of harmony, and for one who can play the violin, there are thousands who cannot." Therefore despite wars hot, cold, and of various temperatures between, despite a world in many ways disordered, a long view does show progress and does give hope for future good. Will Durant defines our position ably in the following words: "I find in nature and history many instances of apparent evil, disorder, cruelty and aimlessness. But I realize I see these with a very limited vision, and that they might appear quite otherwise from a cosmic point of view. How can an infinitesimal part of the universe understand the whole? We are drops of water trying to understand the sea."

We are all inventors in minor things. The one who would improve a thing must realize its present qualities and its possibilities—must recognize with Tagore that the possibility of perfection outweighs the probability of imperfection. Observation is the first requisite, analysis the second, faith the third. Without observation, the need would not be realized. Without analysis, the method could not be devised. Without faith, the impulse would be lacking. The successful effort, then, combines a physical, a mental, and a spiritual activity—in other words, a union of all our available powers directed toward a single goal. To be receptive to the creative impulse, one must have a certain discontent, a confidence in the potential ideal, a sense that betterment is always possible. This gives birth to constructive curiosity.

My story deals less in the mechanical details than in the growth of the ideas themselves—ideas that spring anonymously from the heart of the people, like a folksong that starts, one knows not how, but is played upon by other voices and finally takes shape into a consistent theme. To get mere technicalities, there are already many specialized books. Most of us, however, though interested enormously in the fact that man at one time could not measure the passing hours and now can do so to the fraction of a second, are rather indifferent to the actual mechanics and operation of weight clocks or springs and balance wheels. The development of mechanical perfection is a matter of experiment, but the idea is the point of departure for creativeness.

And where to get the ideas? Sir Joshua Reynolds, the great painter and founder of the Royal Academy, tells us that invention is little more than a new combination of those images which have been previously gathered and deposited in the memory. Nothing can come of nothing. He who has laid up no materials can produce no combinations.

The more extensive, therefore, our acquaintance with

the work of those who have excelled, the more extensive will be our own ingenuity. Then, when the idea comes, we can use it, juggle with it, be receptive to its possibilities—not simply hold it isolated as an amusing or interesting curiosity, but have it as a basis for experiment. Most of us get ideas that we do not develop in this way, and nothing ever comes of them. (The whole subject of "how to get ideas" is developed fully in my book of that title.)

When we talk about ingenuity we may think of the farmer who wanted to sell his farm and called attention to its central location by saying, "You can start from my farm and go anywhere in the world." A real history of the development of ingenuity would lead us equally far. It would embrace the whole story of civilization, and even go back to those prehistoric times when man was adjusting himself to the most primitive conditions. But I shall not dwell upon those ancient days when the Chaldean shepherds first noticed the stars and sought to chart the heavens. I cannot undertake a history of ingenuity in its broad sense —for what is this mother wit but the record of man's adaptation to his surroundings in the most pleasant way? This agreeableness may mean, to some, comfort; to others, idleness; to others, self-defense against an inferiority or an inadequacy. To each person, according to his nature, this sense of happiness will have its special definition; but to all it will mean the achievement of what each one desires, whatever it may be.

Whether it may be true that Adam was happier in Eden, whether civilization as we understand the term has in any essential or fundamental way improved his lot, I do not venture to discuss. One thing is sure: our immortal sense of aspiration constantly urges action, and this action is logically progressive. Many years ago Sir Philip Sidney said, "Who shootes at the midday Sunne, though he be sure he shall never hit the marke, yet as sure he is, he shall shoot higher than who ayms but at a bush." Ever alluring, forever unattainable, gleams perfection. But the measure of a man's success depends as much on how far he aimed as on how far he shot; to fall short of a great aim is often a finer thing than to hit straight into the center of an unimportant one. If we move in harmony with the constructive impulse, we shall go forward the more easily. If our ingenuity carries us in the other direction, we may perhaps take comfort in the thought of the oriental philosophers that we shall have a later opportunity to compensate for our ill-advised actions now.

I shall not discuss the great inventions or the outstanding productions of modern genius, but shall refer rather to those lesser examples of native intelligence that lie in everyone and need only to be recognized to be used. In the commonplace there is much to reward one's interest. The very fact that it has become commonplace suggests an inherent superiority to account for its survival—and its popularity.

Many people, abounding in ingenuity, never make a beginning, because they see only the complexity of the whole model to which they are striving. When a friend called on Michelangelo a second time, he found him still working at the same statue. "You have been idle since I saw you last," he exclaimed.

"No, indeed," answered the sculptor. "I have retouched this part and polished that; I have softened this feature and brought out this muscle; I have given more expression to this lip and more energy to this limb."

"But," said the friend, "these are trifles."

"It may be so," replied Michelangelo, "but trifles make perfection, and perfection is no trifle."

And after all, our world is full of trifles we can try to improve, that may contribute their little quota to the larger betterment. Consider how wonderful is the commonplace! What would you have done, how would you have felt, in the long ago when none of these was known? Surely not like the relatively safe and comfortable person you are today. Man had to create all these now familiar objects to adjust himself to life on earth.

Look at these objects in daily use as if you were seeing them for the first time, with all the awe and wonder they would then inspire. For the commonplace thing is full of romance, of tradition, of interest, whether it be the clock that gives you the hour, the fork you use at table, the shoe you wear, the pen you write with.

In their origin and development is the story of struggle, of achievement, of conquest over obstacles and problems. Today's commonplace is often the sum of millennia of inconvenience and even hardship overcome. It is full of lessons of man's ingenuity and adaptability. The knowledge of it, the observation and study of it, enriches daily life and forms a point of departure for progress and improvement. Yet there are people who take these things for granted, who accept them merely on their face value—their very superficial value—and find them never more than commonplace!

Commonplace? Not if we look beyond the object itself into the story behind it, the traditions, the excitement, the human life and history, the spirit of it. As we look so, the face of the whole world changes. You begin to get out of it the joy and interest you put into it. There is never another moment of boredom. The things are no longer mere utilities: they become friends as well. They have life, force, vitality; they vibrate with that quality which points out our interrelationshp with other people, that causes us to appreciate their contributions to our comfort and well-being, and gives us the glow of friendliness based on mutuality of interest.

II

Counting the Hours

PERHAPS the first truth that Adam discovered was that time flies, and ever since, we have been devising ways and means of measuring it. Today in our leisureless civilization, of which we in America are so proud, Babel itself could not throw us into greater chaos than a single clockless day. Imagine a modern timetable based on observations of the stars. Try to time your soft-boiled egg by the sun. The nearest success, without a clock, that I ever heard of was the old mammy who for a soft-boiled egg would sing one verse of "Rock of Ages," and for a hard-boiled egg two verses. After a while the family acquired an inflexible taste for omelets.

Truly does Emerson call our attention to the fact that civilized man has a fine watch but has lost the skill to tell time by the sun. Today we rely so completely upon clocks that one which registers inaccurately can produce no end of confusion, not to say danger. A variation of a few minutes one way or another may merely make us late for work, or it may make us miss a boat for Europe. It may cause a train wreck, or it may cause a delay in factory production which will mean a loss of thousands of dollars.

Many were the devices, variously ingenious, that had been employed for telling the time before clocks came into existence. At the dawn of history man measured time by the shadows cast upon rocks, a principle later worked out in the sundial. When the shadows were low on the trees or a knoll, this indicated that there was plenty of time left for hunting; but when the shadows had crept up to the top branches or the crest of the hill, that warned the hunter to hurry home and hide in his cave before the nocturnal beasts of prey started forth in their quest for food. Time was, even then, of importance.

Later, when men gathered together in groups about a campfire, it became more so. Tribal life presupposed co-operation between individuals, and this meant keeping appointments, hunting and fighting together, planning in advance for future events. Here it must have occurred to one of our prehistoric ancestors that it would be convenient to have some definite method of marking the moving shadow of the sun. In some open space a pole might be set up that would throw a good, sharp shadow, and a circle of stones could be placed around it. Then, as the shadow traveled around the pole, reaching one stone after another, man could see just how fast the daylight was going. Better still, he could make fairly accurate appointments with other members of the tribe. One would perhaps agree to meet him when the shadow reached the third or the fifth stone, the red stone or the sharp stone. Some such arrangement, constituting man's earliest deliberate attempt to record time, was doubtless the first artificial sundial.

But the sun could not be depended upon for marking distant dates. The moon, however, was more accommodating. It was observed that the moon changed its shape from time to time, always in the same order, looking first like a slender crescent, then growing fuller until it became a perfect sphere. Early man employed this observation of its changing contours to determine the time for his future engagements. So things stood for vast eons, during which primitive man developed very, very slowly from a hunter into a farmer who had learned to plant crops, store food, and domesticate useful animals. Skipping over these countless ages, we come to a period about 4000 B.C., at the dawn of recorded history, when mankind suddenly emerges to our view in a highly civilized state.

The scene is now laid in Mesopotamia, and the people

who are beginning to record their activities are the ancient Chaldeans, or Babylonians, intelligent, shrewd, and progressive. They have cities, with brick houses, shops, roads, laws, a written language, a flourishing trade, and a vast curiosity concerning all the phenomena of nature. The sun and the moon and the stars have now been elevated to the status of gods and goddesses, and the clever Chaldean priests who serve these deities have begun to know a great deal about them. It was these learned priests who first timed the earth in its movement around the sun (though they thought it was the other way about); who first charted the stars in their constellations, establishing the twelve signs of the zodiac; who observed that the waxing and waning of the moon occupied about thirty days, thereby laying the broad foundations of the science of astronomy. With the exception of a few minor changes, they worked out our calendar as we know it today, dividing the year into twelve months, the months into weeks, the weeks into seven days, and the days into hours of sixty minutes. All of these things they did without the aid of any of the astronomical instruments in use today, not even a telescope to aid their own ingenuity.

So long as man led a rather simple life, the fact that a sundial cannot indicate time on cloudy days or at night was of little importance. The time did come when it was felt that this shortcoming should be remedied, and it fell to the lot of some unknown inventor to devise the water clock. Just how it came about that water was put to work to measure time is not known. The idea was hit upon probably because of the ease with which a uniform flow of water can be maintained; and, of course, to obtain a uniform motion for an infinite length of time is the ideal toward which the makers of timepieces have always striven.

The Greeks of ancient days had the custom of placing a certain amount of water in an urn with a slight leak in it through which the water would slowly filter. They had a rule limiting their public speakers to talking one urnful, and to stop when that amount of water had emptied. Thus our modern expression "saying a jugful" is perhaps derived from an honorable tradition. In the course of centuries, improvements and elaborations were made. Charlemagne, in the ninth century, received a very interesting water clock from the King of Persia. The dial was composed of twelve small doors representing the hours. Each door opened at its appointed time and out of it came the same number of little balls, which fell one by one at equal intervals of time on a brass drum, so that one could hear as well as see the hour. None of the water clocks indicated time by a dial and hands, but usually by periodically striking gongs in some ingenious way.

Thus the water clock supplemented the sundial; but inasmuch as water would freeze in winter and evaporate in summer, it was far from satisfactory, so the search went on. The next experiment was the sandglass, which is so familiar that it need not be described. The trouble with this was that someone had to be around to turn it upside down each time the sand was emptied from the top section into the bottom.

Among the other early timepieces was the fire clock, which measured time by the burning away of a stick or candle, usually with notches at regular intervals. Thus when the candle burned to the third notch it was time to feed the baby; we can imagine that at the fifth notch the housewife put the wild boar on to roast for dinner, and at the end notch the family went to bed, which was about all that was left to do when the last light flickered out.

Fire clocks of one kind or another have been used among primitive people in nearly all parts of the globe. King Alfred of England devised a time indicator composed of six candles, each twelve inches long, which burned at the rate of twenty minutes an inch. Thus one candle lasted four hours, and six lasted twenty-four hours. This plan worked as long as the wind did not blow, and to shield the flame from the wind he had cases made of horn scraped very thin, which were slipped over the candles. In Old English such a case was called a lanthorn, and from it we get our word lantern, which might be called a by-product of the search for a clock.

Some Pacific islanders still use a fire clock. On the midrib of a long palm leaf they skewer a number of the oily nuts of a nut tree and light the upper one. As the nuts burn off, one after the other, they mark the passage of time, acting as fairly serviceable clocks.

The derivation of the word "clock" is allied with the Teutonic *Glocke,* the Italian *glocio,* and the French *cloche,* all of which mean "bell"; and it takes no Sherlock Holmes to deduce that early time-telling consisted largely of sounding a bell by hand at regular intervals determined by a sandglass or sundial or fire clock or any of the methods set down. Ships' bells for time-telling are still the accepted method, evolved from this tradition. Striking the hour, then, was one of the first processes, though rather differently performed from our present methods.

The introduction of bells into Christian churches took place about A.D. 400, in Campania, Italy, whence the names "campana" and "campanile," the bell and the belltower. From these the town could hear the time, and the use soon spread over Christendom. Throughout the Middle Ages there was no other source of information about time than the position of the sun, cockcrow, the bellman, and the bells of countless convents and churches ringing for worship. The clergy indicated the hour, but they themselves had no reliable sources for obtaining it.

The logical development from striking the bell was to secure mechanical means of striking; and, with the literal mind of those early days, make-believe men and women were assigned to this undertaking. This is one of the numerous features of such structures as the famous clock of the Strasbourg Cathedral. These mechanical figures were in use before the necessity for dials was felt; and after dials

The Metropolitan Museum of Art, New York

Brass gilt German clock watch, seventeenth-eighteenth century, with one hand, a revolving inner dial, and no glass over the dial.

were introduced, the mechanical strikers continued in favor as well until the modern clock was invented.

The first clock that conforms to our modern understanding is alleged to have been made by an English monk, Peter Lightfoot, about 1335. As if this were not an achievement in itself, clocks immediately were elaborated to indicate also the phases of the moon, the movements of the sun, and to contain various other intricate mechanisms. One of the earliest historic examples of this sort has been described as having in addition a number of small figures which enacted various events of the Resurrection. A cock fluttered its wings and crowed to announce the hours. At the same time an angel opened a door and saluted the Virgin Mary; a dove representing the Holy Ghost descended, a benediction was conferred, a carillon chimed, animals moved their eyes, and all finally retreated as the clock struck.

In 1370 Henry de Vick, a German, solved the problem of the weight clock, whose use, because of size, was restricted to large public buildings. He was brought to Paris to make a clock for the tower of Charles V's palace, and he made one that has become famous as the first public timepiece in that city. The churches were ordered to give the time by this. In a somewhat improved form it can still be seen there in the Palais de Justice.

For the next hundred and fifty years clocks increased in numbers, decreased in size, and improved in accuracy, but without any fundamental change in mechanism. Instead, the craftsman confined his skill to producing beautiful cases and indicating time by automatons rather than the simple dial and hand.

The story is told of a Frenchman named Burdeau, who made an ingenious clock in compliment to Louis XIV, in which Louis was represented sitting on his throne surrounded by all the pomp of royalty, the princes and dukes of Italy, and the electors of the German states. These individual figures advanced toward the king, and, after bowing, would strike the quarters of the hour with their extended canes. There were other figures representing the kings of Europe, who similarly paid their respects. Burdeau was prevailed upon to exhibit his clock publicly; and after he had decided to do so, he made a change in the movement which he thought would be highly pleasing to his king. He knew of the stubborn and unyielding attitude of William III of England toward Louis, and the change in his clock was to cause the figure representing William to make a most extreme bow when he appeared. When the exhibition took place and King William appeared, he bowed very low; but at the same instant something went wrong with the mechanism and King Louis fell out of his throne prostrate at the feet of the British king. The accident was considered a bad omen, and poor Burdeau was sent to the Bastille for his pains.

Up to 1500 the use of clocks was necessarily confined to cathedrals and large buildings because of their heavy weights, but with the invention of the coiled spring and gradual improvement of the works, the general use of clocks extended throughout Europe, at first as the expensive and highly decorated timepieces of royalty and the very rich. Early clockmakers were held in considerable esteem, their work requiring a high degree of mathematical ability, and many of these men were retained at court as "clockmaker to the king."

In 1583 Galileo, then nineteen years old, discovered the mathematical laws of the pendulum in the Duomo at Pisa, and we today are still shown the lantern, almost imperceptibly swinging, from which he made his observations. But these laws were applied to clockmaking only in 1657 by Christian Huygens, a Dutchman of learning who had settled in Paris. The resulting French industry rose quickly and prospered, but was abruptly killed by the revocation of the Edict of Nantes in 1685, for nearly all French clockmakers were Protestants. Many of them fled to Geneva, and Swiss timepieces still possess the prestige acquired at that time.

Portable timepieces could not be constructed until a

compact movement became practicable. The first successful experiment toward this end is attributed to Peter Henlein of Nuremberg at about 1500. Their cost kept them out of the reach of the masses, but wealthy men became immediately interested and delighted in possessing the curious novelties. Some of the early clocks had revolving dials and no hands, the current hour being marked by a fixed pointer.

At first only one hand was used on clocks and watches, and people who were used to no other system were able to estimate the exact time very accurately. Two hands did not meet with general favor for a long while; it was considered a much more complicated process to read the double indication. The minute hand was very gradually acknowledged, but country clockmakers were still producing clocks with only the hour hand until almost the nineteenth century.

Clock hands do not as a rule lend themselves to decoration symbolic of particular subjects, but there are examples of such in the Louvre. Three pairs, typical respectively of sport, agriculture, and music, are known. Each member of a pair was different. The one dedicated to music contained on one hand reproductions of a bow and fiddle and a horn, and its mate bore a harp and a flute. They are dated in the 1700's.

The earliest portable timepieces, which one may be fortunate enough to see in antiquarian collections abroad, were very elaborate in design. Let me describe one of them. The clock has three horizontal band dials showing respectively the hour, the day of the week, and the date of the month. Below are portrayed Adam and Eve in the Garden of Eden. As the hours are struck, Eve turns and presents an apple to Adam, who appears to hesitate, then retires, refusing the gift. Abundance of foliage and fruit is spread over the three trees or columns supporting the dials, while a huge serpent gazes menacingly at Adam. The serpent is twined around the central trunk, his head down, his tail used as an hour hand to indicate the time on the dial.

Many other clocks depicted saints and enacted whole

stories of myth or history. The characters would come out in action when the clock struck, portraying scenes in their past. While many had a religious significance, some were simply fantastic.

Numerous materials were used for clock cases even at this early date. Various woods, precious and semi-precious stones, amber, metal, and leather were all employed.

During this time watches were known, but were hardly distinguishable from small table clocks. The case was a cylindrical box of metal finely chased, and had a lid rising on a hinge to cover the dial. This lid was pierced with an opening over each hour so that the position of the hand might be seen. Most watches were provided with a bell. The plainest cases were of brass or polished steel, but the more costly ones were of silver or gold, or both, further adorned by jewels. Watches also assumed the shape of books, birds, animals, flowers, crosses, skulls, and other objects, all suitably carved or engraved. They were very quaint, beautifully wrought as to craftsmanship albeit often fantastic as to subject. The faces were gold or enameled, until white dials appeared; which was not until the end of the seventeenth century.

Watches when they first came out were, oddly, employed rather for ornament than for utility, for, unlike the reliable timepiece of today, watches then were almost useless. Worn as other jewels were worn, for display, the chief appeal of those early watches lay in their novelty—in the fact that man's ingenuity had scored again. No one expected them to tell the time. The smartest people at the European courts, including Queen Elizabeth herself, changed their watches with their costumes, and some wore two at a time.

Large watches, really small clocks, would be kept on tables or cabinets, and small ones, if worn on the person, were held by a chain around the neck, incorporated in bracelets (Queen Elizabeth's forerunner of the wrist watch), or otherwise attached to the dress. It was not for a hundred years after the invention of the mainspring that watches were carried in pockets. The reason is to be found in the grotesque and uneven cases in which watches continued to be inclosed. With their odd points and bulky shapes, no pocket could endure their company. However, the Puritans, with their aversion to ornament, did conceal them in pockets, and watches then became simpler and better adapted to this location.

Just as the shape of watches changed with the times, so did the subjects of their decorations. Starting with a predominance of religious incidents, case makers reverted in the early part of the eighteenth century to classical lore for inspiration, and later included portraits, rustic scenes, cupids, and a wide variety of topics. Many artists engaged in the decoration of enameled watches and attained results significant for their high degree of beauty. The delicacy and minuteness of detail made many of these of great value, a fact that is readily shown by the marvelous collection of J. P. Morgan displayed at the Metropolitan Museum of Art in New York.

About the same time that the artists were working out their beautiful novelties, something in the direction of efficiency was attempted in the form of an alarm clock guaranteed to wake the soundest sleeper. In this ancient timepiece the rays of the sun, concentrated through a magnifying glass, discharged a small cannon precisely at noon. The brass cannon was firmly mounted on a marble base and was wound up by loading with a spoonful of powder. Above the cannon was placed a burning glass mounted on struts, which could be adjusted to "strike" the noon hour correctly at various months of the year. The touchhole of the cannon consisted of a groove which paralleled the north and south line of the dial, so that at noon the sun would be focused on the powder and make the primitive alarm clock strike. However, if the sun were not out, it would not operate, and on a damp day the powder would not go off. But the waiting one, if philosophical, could always say, "Well, who wants an alarm clock to go off at noon anyway!"

Today we take for granted a glass over the dial of our clocks and watches. It was not until the beginning of the seventeenth century that this feature was introduced. The type of glass varied with the fashion, some being high, rounded ones cut from spheres, some bulls'-eyes with a circular flat center. At the time of which we are writing, keys were used for winding and setting the hands, and were usually wrought with as much beauty and artistic perception as other portions of the clock or watch. Some prideful clockmakers even decorated the backs of the clocks so that they would look equally well from every angle—unlike modern clocks which only show a fine front.

While clocks of modern principle became known in the fourteenth century, it was not until the seventeenth that they were used popularly by the middle classes. In contradistinction to the elaborately ornamented timepieces sported by the rich, there were now available for domestic use clocks that were variously identified by the names lantern, bedpost, or bird-cage clocks. These were either hung against a wall or supported on a bracket. They were wound by pulling down the ends of the ropes opposite to those from which the driving weights were hung. When they first came out they measured about fifteen inches high and five inches square, although smaller ones were produced later on. Brass was the customary material, and the dials were frequently etched very attractively. These dials projected slightly beyond the sides of the frames, and only one hand was used until 1680 or later.

Not until about 1640 did the long-case or grandfather's clock make its appearance, in order to conceal the weights and pendulum which had previously been suspended in full view. The case later became the subject for elaborate ornamentation. In early colonial days in America the "wag on the wall" was one of the quaint clocks. Clockworks were imported from Holland without cases so that space in transportation might be conserved. Sometimes they were encased after their arrival, but just as often they were hung without, and received their name from this fact. The dials were of white enameled wood, brightly ornamented with landscapes, figures, or flowers. The works were wood, and the clock was wound by pulling the end of the chain to which the weight was not attached.

Much has happened since then. The types familiar to us now came gradually into being. Some timepieces have such delicate mechanisms that they are commonly worn on a wristband and some fit into finger rings. Today no makeshifts satisfy us. The Naval Observatory each day at noon sends forth a signal at the instant of twelve. It is broadcast by radio for those who listen in. Clocks are also operated by electricity and are kept accurate automatically.

Scientists are currently working on clocks so accurate that they will not gain or lose a second in a thousand years. They now use atomic clocks to slice seconds into billionths. This fabulous achievement is not merely to delve into the mysteries of time, but has a practical side in a jet age, when two jets approaching each other a mile apart can collide in three seconds.

In this modern age, between the large public clocks and the portable watches (some water-proof, magnetism-proof, and with their own alarms), the considerate person can generally manage not to be late, and, with the facilities for accurate measurement, can budget his time to the best advantage.

Time and its measurement, then, have through the ages evoked man's mother wit and, since they are matters so close to his heart, have been reflected in the writings of many lands. Not only does a great deal of profound philosophy concern itself with the subject, but even forms of literature may be attributed to it. The idea of rhythm in poetry is closely connected with the regularity of passing time. Short stories, too, have been devised to meet the need for haste in the twentieth century.

For all we know to the contrary, an ingenious playing for time in an emergency evolved the serial story. What better example of mother wit have we than that of *Scheherazade*, the clever bride of the Persian king who, believing in the faithlessness of women, was accustomed to killing his wives the morning after the marriage. But *Scheherazade*, we remember, spent her marriage night in telling a story which in the morning reached a point so interesting that the king spared her and asked the next night for the sequel. This went on for a thousand nights until a son was born. The king admired her ingenuity, loved her, and generously spared her life.

With all our timepieces, most of us have but a sketchy sense of time and timing. The amount we waste of it in trivial, not to say harmful, ways is deplorable, since, once lost, it can never be regained. And haven't you seen people rushing when there is ample time, while others dawdle when they should accelerate?

There is more to be said of an hour than that it has sixty minutes. In addition to its duration, there is its intensity. One hour, if enjoyed, can seem like a few minutes, while another, saturated with fear or pain, seems far longer than its sixty minutes. No timepiece has yet been invented to measure these individual differences dependent on the emotional intensity of the experience within any particular hour.

How true it is that "He gave unto man the understanding of the shortness of time." This is but one of the limitations that come to us when we deal with the infinite that we can only partly understand and partly represent in our finite characters. All infinity may be for us, but we can use only the present moment. It is by using it to full capacity, with our whole souls, that we enrich our lives and make them the deep, vital, full creations that we sometimes see in our visions. Time is perhaps our one leveler. In this great gift we are all equal. Without it we are nothing. With it we are all that we can be. The instruments for measuring it are verily results of ingenuity, but these are only the means to the end.

In this era of Einstein, perhaps we should beware of

clocks and calendars and all those limiting instruments of time. Perhaps with all their accuracy, they may give us a false notion, imparting unjustified confidence in any apparent progress, or needless discouragement at its absence. We like to hear clocks tick or strike; we like the finality, perhaps, of the roundly numbered days and decades. But the real day is a responsibility and an opportunity rather than a date. Actual time and actual space, we are learning to understand, are what they are only relative to the surrounding conditions.

A story is told of the great Roger Bacon, who by following some ancient writings, after many years of experiment, made a man of brass. It was promised that the "man" would be able to reveal a most valuable secret. When the great test came, the man of brass whispered, "Time is." A moment later his eyes glowed, and more loudly and warningly he called out, "Time was." Finally he thundered, "Time is passed." With that he fell into a thousand pieces. Bacon was so grieved at this wreck of his hopes that it is said he burned his books and retired to a monastery. But after all, while the "secret" may seem obvious, it is a great thing to realize that time is now.

In our age of speed we are sometimes asked, "What do you do with the hours you save by taking the fastest jet?" Well, it is certainly true that if we waste them it is not the fault of the airplane. When there is nothing apparently to *do* with such odd moments, we might pause to *be*—to use them for thoughts of beauty or peace that in a busy and crowded world will bring repose and refreshment.

It is never too late. If there is anything we know, it is that the next minute will have sixty full blessed seconds like all those that have gone before. Whatever we really want to do, the best way to begin is to begin.

One of the profounder influences on our life is the great trend toward automation. Though we admit the transitional difficulties, it is bound to create immensely more leisure and we shall be counting the hours in a very different way from the way we do now. If only a fraction of what technologists promise for the future comes to pass, within a very few years automation can and should make possible a four-hour work day, a four-day work week, longer vacation periods, opportunities for earlier retirement, and the like.

Perhaps by hurrying along as fast as possible to the goal of physical happiness, comfort, efficiency, or what you will, we shall the sooner be able to give more time and consideration to the more thoughtful matters—the invisible forces about us from which we draw our creative power or ingenuity in the first place.

Meanwhile, let us have pleasure and confidence in man's ascending effort. With man's ability to foretell eclipses to the second, ages in advance, we can realize we live in a universe founded on a balance of wisdom, power, and love, and if we cultivate these virtues as they are intended, we can look forward hopefully with the feeling that "the best is yet to be."

Time is fleeting, the clock reminds us, but Time Is, eternally; and however fast it flees, there is ever more of it. It is the inexhaustible gift of manna presented fresh each day to all alike. It is here for us now. The instrument for marking it helps us to plan it effectively and advantageously, but the thing itself is the very basis of all that is.

III

The Struggle for Fire

IMAGINE yourself in the guise of primitive man untold centuries ago. For the very first time you see fire. Things look different, lighter, brighter. Warm, pleasant sensations assail you. But a sudden wind drives the flames along and they ruin a grove of trees. Then your pet tiger dashes into the fire and disappears as a shrieking red devil. A bird is sucked into the flame, and it smells and tastes like nothing on earth ever did. Where things were green, they now are black; your eyes tingle and everything is queer. Of all the bewildering series of events to confuse primitive man! Puzzles, mysteries, possible new powers, new fears, new wonders, a challenge to all his wits and brains.

Primitive man regarded fire as one of his most sacred possessions, its very advent upon earth having been veiled in the unknown. It was to him as a strange message from an unseen Being, full of wrath at one moment, distributing blessings the next. A terrible and wonderful thing of mystic origin, impossible now to do without, and most difficult to reproduce. No wonder that he guarded it above all else.

Fire was a discovery. Man had to find it accidently, for he was unable to imagine a force that could do what this could do. Forest fires started by lightning, volcanic action, desert heat acting on dry, combustible substances, strange chemical combustions—all might serve to produce fire in various parts of the world.

Fire is not a subtle thing. It shows with perfect frankness what it can do. There is no hesitancy about it. Man, while terrified at it, also felt its warmth, saw its light, and gradually realized its other powers. By some instinct he knew that he had to be able to command it, if he were not to be at its mercy. In those days, to produce it at will was the great problem.

When nature made fire, man had to transport it by firebrands which he carried from the source to his home. Naturally, he protected it as long as he could, but a gust of wind or a rainstorm did to fire then what they still do; and it was impossible to keep a fire very long. Another journey, a dangerous, difficult one perhaps, to a volcano or to the edge of a forest fire, would have to be made!

But man, once learning the use of fire, could hardly wait for a volcano to go into action, or for the special conditions that were required. He started to think. And by observing, and analyzing, and experimenting, his mother wit enabled him to devise ways for making fire when he should want it. The invention of means to make fire by friction was an act of the intellect which in its own degree demanded as much thinking power as the invention of the steam engine. Percussion was also used later. After many centuries, the burning glass was sometimes employed to focus the sun's heat upon a substance that would burn readily. Ages later, again, the tinderbox would be brought out and a light made from flint, steel, and tinder with the aid of a stick with sulphur on the end of it. This was called a "spunk," grandfather to the match. Our "spunky," meaning to fire up quickly, comes from this word.

What a long way to the development of a match, the little insignificant match that we use so nonchalantly at the rate of billions a week! This little instrument first brought chemistry consciously to the aid of fire-making about three hundred years ago, and many improvements were necessary before the safety match and match folder of today were evolved.

The use of coal was not known to the ancients; nor have we found out when it began to be used for fuel. This was in England, and it was well known there in about 850.

From the very beginning, fire has had countless uses, but man alone has been able to create or defy it. All the talents of all the lower animals stop short of that. Indeed, man has used it in his own interests and against theirs. In hunting, for instance, if he wanted to drive a herd of animals in a certain direction, he could start a fire behind them and frighten them along by cutting off any other road of escape.

Many pursuits, even primitive ones, would have been impossible without fire. For metalwork, pottery-making, and cooking it is essential. Cooking was, of course, of the simplest. Its first form was roasting. I am sure most of us remember the inimitable story that Charles Lamb tells us of the origin of roast pig. A careless boy accidently set fire to the little hut in which some new-born pigs were quartered. The house burned down, but that was a mere matter of some new straw and a couple of hours' work to replace. But the pigs! He felt of one to see if it might still be alive! He burned his finger and put it into his mouth. A wonderful flavor assailed his nostrils. What a rare taste! Soon he was so busy eating the roast pigs that he did not even notice his father's beatings. In the same way the father, lamenting over his pigs, touched one, burned himself, and put his finger to his mouth. The same result.

Soon the neighbors noticed that one conflagration after another was taking place on this man's estate. They watched him, discovered his secret, had him arraigned in court, and the court, sampling some of the pig, reported "not guilty." As a result, people began building frailer and frailer huts, and burning them down oftener and oftener, until some ingenious person discovered that it was not necessary to burn down the house to burn the pig!

Whether we want to take this story literally or not, we may be quite sure that some such accident was at the basis of the discovery of cooking. When a man found that fire improved the quality of his food, he tried to kindle a fire on the floor of his cave. Though it filled with smoke, he did not mind. There was the warmth, the better eating, the light that made the cave useful as a place to do things in. Later he made a hole in the roof of the cave or the hut and let the smoke escape. That was the first stove. It was the whole house. Man literally lived in his stove.

Around this fire were grouped the rude cooking utensils, for in those days the fireplace meant hot food and bodily warmth, and romantic daydreaming was a mere by-product of an intensely practical household device.

Perhaps the greatest good that fire brought to man, if we except the illumination of his imagination, was the development of the instinct to create a home. *Wanderlust* is not an instinct inherited from nomadic ancestry. In early days men were nomadic because they were compelled to follow the roving beasts upon whom they were dependent for supplies. But nothing less vital, less stimulating, less incalculably valuable than fire could have overcome that long-established custom.

Today the idea of a family is a familiar one. It is difficult to say how far family life was developed by the fact of fire. Duties of the family increased because fire had to be cared for, protected from going out, from burning the babies and doing other damage. Perhaps the first person who had a steady job was the one who tended the fire.

The fire drew people together; they all wanted the benefit of its warmth and light and protection; they built the hut to care for it; they grouped themselves around it and developed ways of living in amity and coöperation.

Someone has remarked, "We cut a tree down to burn as firewood to warm us—but what do we do while we are warm?" Is there any sense in destroying a majestic tree to warm us so that we can dissipate our time and energy over stupidities?

Beginning with the ancestral cave man, when the fire was the only source of bodily warmth and hot food, it has been so intimate a part of domesticity that the hearthstone has been figuratively thought of as the home itself. Almost all races, in fact, have worshiped fire in one form or another, and many religious cults have sprung up concerning it. From most ancient of days and still in the present there are tribes that have certain fire rites. Do we not remember Vulcan, the god of fire and the forge, who protects all artisans and whispers inspiration in their ears as they work with him?

In days when it was so difficult to produce fire, a flame would be kept constantly burning in some public place, from which private fires might be kindled. In the old mythology, Vesta was the goddess of the hearth, and fire was in her gift. Because she scorned her earthly lovers and vowed herself to eternal chastity, she was given a seat at every man's hearth to be forever a symbol of purity and light. The Roman Forum, as late as Caesar's time, had Vestal Virgins to take charge of these sacred fires, and to permit one to die out was to invite being burned alive. This indeed is the theme of Spontini's brilliant opera, *La Vestale.*

Not only in classic times but in medieval Europe during the twelfth century the fireplace occupied the position of honor. Here it was in the center of the room. The fire was kindled on a rough stone altar and fed with wood logs and peat, while the smoke escaped through an unprotected hole in the ceiling which also let in the rain. Much of the smoke settled on the beams as soot. In damp weather the soot absorbed so much moisture that it became too heavy to stick and came tumbling down in unwelcome ways and places. While no one would call this pleasant, it did have the effect of somewhat modifying the foulness of the floor, which was covered with rushes or straw.

In the baronial castles maintained by the nobility, the fireplace occupied a great hall which served both as kitchen and reception room, for the lord of the manor and his retainers gathered about the fire to watch the preparation of the meal. The high seat was the place of greatest comfort opposite the fire, which of course served not only to cook the food but to light and warm the hall as well. To make the fire burn more brightly, it was later placed at the end of the hall, where the door would furnish the required draft. The middle of the long seat was no longer the place of greatest comfort, and the master shifted the seat of honor to the end of the long table instead. This had its bearing on subsequent etiquette, which places the host's seat in this position.

The transition of the fireplace from the center of the room to the end of the hall logically developed into a desire for a separate kitchen. This must have been created for the comfort of the rich, who thought they would prefer an element of surprise in their feast and not have it prepared in full view. An old Icelandic law provides that if there be both a hall and a separate seething house, or kitchen, the householder must declare for which of the two the community shall be responsible in case it is burned down. This shows the existence of a system of communal fire insurance at a much earlier date than we might have expected.

By the time all this had developed, coal was a major fuel. It seems not only one of the most important of minerals, but its whole story is one so filled with romance that it is almost a fairy tale. We call it a mineral, and we mine it; yet it is formed of vegetable matter ages old and created by the slowest evolution. All the coal now existing was formed untold ages ago when the conditions of temperature, moisture, and atmosphere on the earth's surface were different from what we have today. Coal is not renewable, then, and we are living rather on the principal than on the interest of the available supply.

Though ancient races developed fine civilizations without coal, these were chiefly peoples of warm countries. Since the general use of coal, a whole new form of civilization has been created, and industries and methods adopted that have extended and expanded the interests and activities of this life almost unbelievably.

Coal or material produced from coal has been used for almost all domestic and industrial heating, and as the principal source of power for railroads and steamships, until the relatively recent age of petroleum, and has been indispensable in the manufacture of iron and steel. But more interesting by far than this are the hundreds of useful by-products derived from coal, from which ingenious modern chemists have developed multitudes of new substances of great value to present standards of living. Such diverse things as fertilizers, explosives, food preservatives, disinfectants, antiseptics, perfumes, photographic materials, roofing compounds, road materials, medicines, paints and varnish, flavoring compounds, and dyes may be mentioned. If we read this list thoughtfully, we can see that this magic product has made inroads into almost every necessity and luxury. Today the same things can be said of petroleum—its prehistoric age, its non-renewability, its uses and its by-products. To take but one use, were it not for the lubricating power of petroleum products, the machine age itself would be impossible, for the friction of moving parts unoiled would grind them to a rapid halt.

If man had not had fire and had not been ever striving for new ways of obtaining it, who shall say how many of these things would remain unknown? And now our mother wit should enable us to insist upon ways whereby coal should first be treated for the recovery of these valuable by-products, allowing only the residue to be burned for fuel instead of the raw coal; and that, wherever possible, water power and fuel oils be used for mechanical purposes, saving coal itself for those exclusive properties for which there are as yet no substitutes. Anthracite alone does not produce much of value in the way of by-products, and is a proper fuel for direct burning. It is chiefly the bituminous and other soft coals which have these further uses. The soft coal that was used in medieval times gave off so much smoke with its unrecognized by-products that in self-defense the chimney was finally evolved.

The chimney had its origin in the fact that in medieval England much fighting was done on the roofs of the fortress-houses. The smoke coming up through a hole in the center of the roof both caused the soldiers to be very uncomfortable and interfered with their efforts. That encouraged the moving of the fire to a place near an outside wall, and an opening was made in the wall above it; in other words, here was a chimney. The chimney was unknown to the ancients. Though the Bible speaks of smoke out of a chimney, the word here translated as "chimney" meant simply an open grating, possibly in the roof, for the escape of smoke and the admission of light. The Hebrews and the Greeks had no such thing as a chimney as we understand the term. The most that Roman luxury required was to heat air in one room and convey it by pipes into another. This was a very satisfactory system of central heating which served all the purposes of the modern furnace, which has only the pipe to the roof to differentiate it.

The coming of the chimney transformed the hall. Since the smoke could now pass directly and entirely into the open air, the space for it overhead was no longer required. First, part of this was made useful by building an upper floor, which gave a loft extending over half the hall. Soon the loft reached over the other half as well, and the house had a second story.

In the fourteenth century still more privacy was obtained by the addition of another room at the dais end of the hall. Being smaller, it could be made more comfortable than that damp, chilled setting of the general hall. But it continued to be true that anyone standing or sitting near the fireplace was nearly roasted on one side and half frozen on the other, in spite of all that high-backed settles could do. We do not begin to realize the drafts to which our ancestors were constantly subjected. A man needed his heaviest furs less for outdoors than for sitting at home. No wonder so many of us have had an inherent dislike of drafts; it must be derived from the memory of past unpleasant experiences.

To save fuel in heating the halls, there was devised a

structure of masonry with space for a fire inside and at the top an escape for flame and smoke directly into the hall. Into this the fuel was introduced every morning, and a brisk fire secured to warm the whole structure, which then radiated heat all day. This was the ancestor of the earthenware stove of the Scandinavian countries, Russia, and Germany.

Great tile stoves have long been popular in the cold countries of Europe. Germany, Holland, and Switzerland use them even today. A stove like this is really a monumental feature and gives its name to the room, which is called a *Stube*, after this valuable stove which makes it livable. Such a heater is generally located at the center of the story, adjacent to the kitchen, and connects with the central chimney. The rooms are built around it, as many as possible benefiting. Ornate stoves bear Bible history in each of the panels, of which there are many, adorned by borders and other designs, and setting forth the story in decorative script as well as picture. Others, simpler, bear dates, coats of arms, inscriptions, and plainer devices. The furnace where the fire is made is outside in the hall. These stoves, when heated, will retain their warmth for many hours. Usually a pair of high steps upon which one may sit will lead to the flat roof of the stove, which is frequently concealed by some device of tile, wood, or curtain. Behind this screen in simple homes may hang the latest wash or drying fruits. Sometimes on the coldest days the children scramble up and find real comfort lying on the top of this great heater. Frequently these stove steps may lead to the upper story by means of a trap door which also lets the heat out into the room above.

For keeping the hands and feet warm, small devices were easily improvised. My own father used to tell of days long ago when as a little lad he trudged to school in frigid weather with a hot hard-boiled egg in each hand as a hand-warmer, which later in the day served for luncheon. In Holland I happened into a little old church in one corner of which was a pile of small wooden boxes about six inches high and eight inches square, with a hinged door at one end and a perforated top. We asked what these were and were told they were *stoofjes,* or foot-warmers, for the tile or stone floors are very cold in winter and such protection is necessary. Most private dwellings, I noticed later, also have these. In warm countries like Spain or Italy, where just a simple expedient is necessary to take off the chill from a room, braziers were used—little trays of charcoal placed on a pedestal. In colder places, heated bricks or stones, warming pans, and other things were obvious devices.

Even in late colonial times it was the task of many a lad on many a bitter morning to trudge half a mile or so in the biting cold to some neighbor in order to borrow a fire. This was brought back smoldering low so that it might be ignited in a full blaze on his return. Careful he was of it, too, for he knew that if it went out before he arrived, he would have to go back for another. Such an ordeal was the making of fire that often it would be kept burning all night, at the cost of large quantities of wood, just to save the effort of starting it up in the morning, or sending out for it.

There is a wealth of romance in the old colonial fireplace, where so many converging interests met. One can almost think back to the kitchen tools dangling over the fire, the drying pumpkins, the red peppers, the apples and other goodies close at hand, the steaming pots and kettles hung from a crane, the aroma of roastings and bakings, and all the rest. In those days the kitchen was the living room of the house, because of the comfort of the fireplace; and there, if there were an idle moment, the thrifty woman would try to catch up on her knitting or spinning.

One of the nicest fireplaces was the one arranged within an inglenook. Inviting fireside seats separated it somewhat from the rest of the room, shielding those about the fire from any possible drafts and adding to both their comfort and their privacy without depriving the rest of the room of glimpses of the ruddy light and warmth. Such an inglenook was contrived early in the days of fireplaces when they were moved from their first position in the center of the room to a recess in a side wall. Those who tended the fire and the roasting spit naturally stood on the hearth within the walls and under the hood. It was logical to place a settle against the inner side wall, or indeed a bench at either side. This location had its decided attraction on cold nights and came to be the place where the elders sat, entertaining guests, instructing their young people, gossiping with neighbors, or discussing their own private affairs. Outside Stratford-on-Avon in the adorable home of Anne Hathaway at Shottery, I saw just such a nook where Shakespeare courted Anne under the vigilant eyes of the rest of the family.

Regrettably, in our material advancement we have felt of late years that a stove is more efficient, a furnace makes less trouble, a steam system is cleaner than a hearth. Fuel oil, radiant heat to make us warm, air-conditioning to make us cool, have entered the scene, with the result that the old-time fireplace has dwindled in importance. It is true that the beauty of a real open fireplace may be rather successfully improvised by the use of artificial electric coals, and by logs that are not of wood but of asbestos simulating wood. Both of these give a very genuine-looking flicker that satisfies the imagination infinitely better than no fire at all. These make it possible to have an active hearth without work or annoyance, and eliminate a little of the danger of an open fire. But the comradeship of fireside friends, the restfulness of the open hearth so needed to offset the intensity of modern life, have been too much ignored. Who that has known these can rise to the charm of a radiator, or glow with poetic reminiscence before a hot-air register? Shall the loved phrase "For hearth and home" be superseded by one that says, "For stove and home" or "For radiator and home"? Can we make fairy tales and twilight confidences, can we admit Santa Claus, through a stovepipe?

Fire is the master magician who touches plain wood or coal with a wand of life, releasing its imprisoned spirits of light and heat. It also somehow liberates our imagination, opens, as it were, a secret door within us, so that we dream and plan and expand into temporary poets and philosophers, sound depths and soar heights never explored except under its influence. Truly, romance must be thin indeed without a fireplace, for the open fire is more than a source of heat or efficient ventilation; it is a link with nature and with beauty, with tradition and with history. It is, in an age of mechanics, a grand survival of the noble, striving spirit of other days. Heat without visible flame may warm a person, but it does not touch the heart.

IV

Dispelling the Night

LIFE in its simpler forms had only a campfire and a torch to prolong the dying light of day. When we stop to consider what our evenings would be with only these, it at once becomes apparent that we would have to revert to the primitive existence of barbarians. We would have to eliminate all activity after sundown, shrink our day to about twelve hours instead of seventeen or eighteen, and cut off about a third of life in so doing. Thus, while night light does not necessarily make us live longer as to years, it enables us to live more widely and deeply.

The progress of lighting has been a very slow one; it is only in recent years that lighting comfort could be taken for granted. But there were so many forces to conquer that having any light at all, no matter how ineffective, was always a blessing.

We have speculated in the previous chapter how fire first made itself known. We can imagine primitive man watching it, absorbed, entranced, as we are even today. How much more so would he be when the experience was utterly new and without precedent! As that memorable night fell, he continued to watch the fire, not noticing the dusk. Suddenly he found it dark, yet light . . . light in a new way—light by means of fire!

Time passed. Then some genius of that prehistoric time snatched a firebrand from the flames and used it to light his way in the darkness. Thus came into being the torch, which for centuries remained the sole means of night lighting. The light of sun or moon, the flare of the volcano, the flash of lightning man accepted. He could not change them. But now . . . here was something he could achieve for himself, when and where he wanted it! The thrill he had of this must have been beyond our most vivid conception.

Much as man doubtless enjoyed the warmth of fire, it is probable that its lighting quality was even more important. People unaccustomed to domestic comfort, nomads, and with little in the way of clothes, were hardened to climatic conditions. Hot or cold, it made not so much difference. But there was naturally a fear of the night. Then came prowling creatures, weird sounds, unseen things of mystery. The conquest of night must have been hailed as a real achievement. Here were real protection and companionship. One could know what was going on, and one could do things, too, even if the pale moon were hidden.

Our earliest ancestors were in no sense nocturnal creatures. The period of darkness was to them the time for sleeping, and their waking hours were those between dawn and dusk. Only when man had reached a relatively high plane above the other members of the animal kingdom, therefore, would he think of protracting the day or dispelling the night.

The tribal fire then, with its firebrand, was followed by a torch made of slips of bark and wood tied together by vine tendrils, dipped in animal fat, and held in the hand by the lower end. The next step was to attach this to a cup to catch the hot fat, which all too often spoiled the temper of the one holding the torch. Torches of this general kind were used even as late as Homeric times in Greece.

We do not know just when the ancient lamp was invented in the form of a vessel filled with oil into which some kind of wick was dipped, but its invention certainly antedated the present era by several centuries. And once this smoky, foul-smelling lamp had been discovered, it remained in use, practically without change or improvement, until the end of the twelfth century, the date of the invention of the modern candle. By 400 B.C., lamps consisting of a terra-cotta reservoir and a spout for the wick had come into general domestic use. These small, flat oil lamps, found in the Nile Valley and later adopted in Greece and Rome, were the chief resource of the "illuminating engineers" of ancient days, and were entirely unsuited to large-scale activity after sundown. Herodotus, instead of using the word "night," employs the picturesque and significant phrase "the time of lamps."

In ancient days there was much reference to the inconveniences and shortcomings of the antique lamps. The parable of the five foolish virgins is perhaps the classic example drawn in realistic touches from the life of the times. These lamps were small, giving a very meager and uncertain light which the merest breeze could immediately extinguish. Crude as they were, the shallow terra-cotta lamps with their smoky flame and dingy light represented a great step in advancement over the torch.

Probably the most wonderful lamp of antiquity was the one of gold made by the sculptor Callimachus for a temple in Athens. It was a huge affair and needed filling but once a year. Its great size and beauty made it one of the marvels of the ancient world. But the Romans held the little terra-cotta hand lamps in highest favor, extending their use for domestic purposes and for the lighting of shops and bazaars. They gave them almost universally for birthday gifts, using the symbolism of light to express their respect.

Nowadays, when even gas jets and kerosene lamps are considered obsolete, that ineffective form of illuminant, the candle, seems about the most primitive type of lighting equipment. Oddly, however, the modern candle was a relatively late invention. Perhaps this was due to the fact that tallow was its chief ingredient, and this was only available after cattle were domesticated. Even then, the fat was also good to eat in suet pudding, and food was not so plentiful that this important use could be minimized. Then, too, in warm climates suet or beef fat became too soft to be satisfactory in candle form.

The first candles to be used were rushes dipped in tallow or oil. The distinction between oil lamps and candles in those days was small; the main difference seems to have been that oil lamps used liquid fuel, whereas candles were covered with solidified fat. For ancient palaces and temples, candles did exist that were made of wax. Beeswax was found excellent and became an article of ancient commerce, but it was not plentiful and was therefore costly. Because it was so laboriously gathered by the bees, southern Europeans considered it as sacred. Therefore wax candles came to be reserved for ceremonial uses such as weddings, funerals, and other temple functions. In ancient literature such candles are frequently mentioned. The Bible and Greek and Roman writings are full of references to them and to the candlesticks, but they were not used domestically until much later. Oftentimes a symbolic meaning was wrought in the figurative language, showing that light and its agencies have enjoyed from the dawn of history the highest admiration of mankind.

After many centuries wax candles gradually came into general use among the aristocracy and the upper classes. While they were expensive, they gave a better light and produced less offensive odor. These ushered in what has been termed the golden age of candlesticks, the seventeenth and eighteenth centuries. Workers in the precious metals made them of rare beauty and symmetry. But the short and stubby or tall, slender tapers which they contained were almost as valuable at one time as the holder itself. Such was their worth in the France of Louis XIV that the ladies of the bedchamber supplemented their slender salaries by "scavenging" the remains of tallow left in the sconces after a brilliant court celebration, and this in the time of one of the most extravagant dynasties the world had known.

In Colonial America candles were not a luxury or an aspect of interior decoration; they were the only available source of light after dusk. Part of every housewife's duty was the dipping of candles, and every attic or workroom had its tallow vats and candle racks. Benjamin Franklin must have had a hard time with his lighting arrangements in those days. A candlemaker as a boy, he undoubtedly had the best in candles as a man. But he drew lightning from the skies to popularize electricity. He invented bifocal eyeglasses to improve his vision in ill-lighted days. He had to practice "early to bed and early to rise," to make the most of the natural light. How he would have enjoyed the lighting of the twentieth century! He would almost have doubled his working time and the number of benefits he conferred upon a grateful nation. Well, we at least have the lighting. Perhaps we can use it as a point of departure for more constructive effort.

Candles or tapers are still used in exploring some of the catacombs in Italy. I remember that any interest I might have taken in the gloomy labyrinthian ways of the catacombs of St. Calixtus in Rome was entirely lost in discomfort at the imminent probability of having candle grease dropped upon me by the monk who followed right behind me. Fourteen hundred years ago that would have been expected, but today a flashlight, while not nearly so "atmospheric," would have contributed a counterbalancing sense of comfort and safety.

Perhaps the desire for steadier and safer light caused ingenious man to develop the idea of protecting the flame from the breeze. He made lamps of bronze and designed a cover to protect the light. These were in effect lanterns, and were considered very wonderful. Lanterns were not, however, difficult to invent. There was needed only a cage for the lamp or later for the candle, and the cage required holes through which the light could shine, or a transparent covering like the paper used by orientals, or the horn plates used in Europe. As far back as 5000 B.C., lanterns were known to the Chinese. They were generally of gaudy coloring on rich silk. Frames were of gilded wood or bronze, and long silk tassels hung from them.

Many picturesque stories have been handed down about

lanterns. The origin of the Festival of Lanterns held annually in China is ascribed to a popular legend. It seems that the beautiful daughter of a nobleman leaned too far over a pond one evening and fell into the water. An immediate search was instituted, but she could not be found because of the darkness. The nobleman ordered lanterns, by whose light the girl was finally rescued. Her father was so thankful for her safe recovery that he ordered the Festival of Lanterns, a custom which has been adhered to ever since, and which is the scene each year of brilliantly lighted and gay processions, when thousands of lanterns line the streets and are borne aloft by the people.

There is also the recollection of Diogenes, the Greek Cynic, who liked to say when he was seen with a lantern that he was looking for an honest man.

Lanterns were used to light the romantic Venetian canals in the height of their glory, and were placed on the corners of important buildings throughout Europe, to give a dim glimmer in the night. Two world-famous lanterns of the sixteenth century were erected on the Strozzi Palace in Florence, where they may still be seen. They are of wrought iron, and real works of art.

And coming down to modern times, there was our old friend Paul Revere, who ordered two lanterns hung aloft in the belfry of the old North Church to warn the countryside that the British were coming by sea.

Modern lanterns used today in halls and corridors are very similar to the old lanterns of iron carried by watchmen a couple of centuries ago, but in the modern fixtures glass has been substituted for horn, and tungsten lamps for the tallow candles or the smoky oil wick.

During the Middle Ages, Europe initiated almost no advances in lighting over the ancients. The great feudal castles, which by day were gloomy enough, were far worse after nightfall. The only means of illumination were the smoky and foul-smelling tallow candles and the flambeaux or torches that were employed to provide greater illumination for banquets and other social events. These gave a minimum of light and a maximum of smell, and were not only extremely disagreeable but ineffectual.

Fires flourished, many of them making great havoc; there must have been many unheard of, with less dire results than the one of which, for instance, Sir Isaac Newton was a victim. After the publication of his book setting forth the law of gravitation, Newton was working at chemistry, but all his labors were lost by an unfortunate accident. One morning he went to church leaving a lighted candle among the papers on his desk. It is said that his pet dog, Diamond, upset the candle. When the astronomer came home he found all his papers destroyed. His health was impaired by the accident, and though only fifty years old, he made no new great discoveries in the last thirty-four years of his life. This tragedy due to the imperfections of candlelighting took place as late as 1687.

For a long time during the Middle Ages artificial lighting was largely dispensed with. One depended merely on the hearth fire, and only supplemented it by a candle to light the way to bed. But the time came when the hearth fire was succeeded by the porcelain stove; and when the light of the hearth was no longer available, lamps were used to light the hall. These took the form of bowls of fish oil, with vegetable wicks afloat in them and burning. They furnished more smell than light, but the world had long to wait for a lamp that proved satisfactory. Only in the eighteenth century did a Frenchman, experimenting with a glass tube held over a lamp flame, find that he could regulate the supply of air so as to obtain a practically smokeless flame. His name was Argand, and Argand lamps were born.

Until then, the nightly gloom still cast itself over the cheerless sleeping chambers, and because there was no way to produce a light at the snap of a switch, primitive lamps were kept burning all night long.

In France in the seventeenth century the *torchère*, a cluster of candles arranged on a tall standard, was hailed as the means of revolutionizing interior lighting; it gained great vogue among wealthy people. At that time it offered a new way of mounting a candle and could serve purposes that a candlestick or a candle stand placed upon a table could not serve. But we can nevertheless imagine that this contrivance, although it added considerably to the flexibility in lighting that up to that time had existed, still left a great deal to be desired. The flickering lights and shadows cast upon the walls of a lonely castle as each gust of wind entered through the cracks must have given a weird effect, to say nothing of the weak, trembling, and uncertain illumination there would be. In later years, when for one candle two or three were substituted, the result was regarded as close to perfection.

The development of *torchère* lighting, however, received a setback because of the fact that experimenters were working with lamps. Glass chimneys were devised for lamps that burned vegetable and animal oils. These early lamps were not particularly successful until a great deal of experimenting had been done in the designing of suitable burners, wicks, and glass chimneys, and extensive testing for suitable luminating fats and oils.

The best of the oils tried out proved to be whale oil, and in America a great industry grew up to supply the demand. Whaling was very profitable, as much as $4,000 being paid for the oil obtained from the blubber of one whale. Whole fleets of whaling ships set out in the early 1800's from such New England ports as New Bedford and Gloucester. This one development carried with it a marvelous amount of romance which inspired such writers as Herman Melville to give us their fascinating stories of the sea and the early American merchant marine. Mineral oils began to take the place of whale oil about 1859, and each year saw a wider use of petroleum products, until they shortly superseded whale oil entirely.

In the early nineteenth century arose an important step in advance of all previous lighting expedients. This was the discovery and widespread use of illuminating gas. First investigated in Glasgow, centers were soon established on two continents for its exploitation. At the outset, suffocations, explosions, asphyxiations, and other ills occurred. It was not easy to adjust to gas. But finally the housewife could bid farewell to the candle grease that had dropped on clothing, stained the carpets, spotted the furniture, disflavored the food, and made itself a general plague—to say nothing of the meagerness of the light that could not but strain the eyes! Now, however, it seemed that the world was ready for real light, and quickly on the heels of gas came the development of electricity.

Electric lights are so familiar today that it is all but impossible for us to realize what an amazing source of wonder they were when first installed on a trial basis, and that only a comparatively few years ago. "It seemed almost like writing by daylight to have a light without a particle of flicker and with scarcely any heat to make the head ache," said *The New York Times.*

We do not always recognize the invaluable contribution of glass to the art of successful lighting. Without it the use of electricity for this purpose would be impossible. Lighting previous to the use of glass was attended by grave dangers, horrid odors, smoke, and soot and, for all that, was anything but satisfactory as light. It was dim, unsteady, and apt to go out at just the wrong moment, or perhaps to explode.

There is a tremendous contrast between the little guttering, darkling lamp of ancient times and the brilliant bulbs used to illuminate our white-light districts now. The throwing of a switch, the pressing of a button, gives a result that not so long ago might have been regarded as some sort of supernatural magic or witchcraft. Today, however, we are growing accustomed to the fact that the universe is full of invisible forces, like electricity, that we do not yet understand, and we eagerly await the new discoveries as they occur, holding an open mind as to their possibilities.

The coming of gaslight and electricity not only greatly improved interior illumination but made possible the policing of the towns and the increased safety to property and life. The modern officer on his beat succeeded the old watchman carrying his horn lanthorn, calling out the hours and preserving the peace. In fact, it was not until the middle and late 1800's, after the coming of gas, that street-lighting became at all common, and the picturesque but humble and weary figure of the lamplighter came into his own. Before that, street-lighting was not a public service or enterprise, but was done entirely by individuals sufficiently adventurous to be abroad at night. People who were brave enough to stir away from their own doorsills went as a rule either by litter or by sedan chair, preceded and in many cases followed by their own torchbearers. This practice continued until quite recent times. It is said that there are still in existence, in certain old cities of Europe, houses that have at their entrances the racks used to hold the torches until the visitors decided to return home.

In the London of Queen Anne's time it was entirely unsafe to trust to the feeble illumination from house windows. A ruthless class of vagabonds and criminals operated at night, and a journey across the city was a hazardous adventure. Little boys could be called to light people to their destinations, just as we today might call a taxicab. These boys were known as "linkboys," and for a few pence performed a real service. Occasionally smoky oil lamps were hung on ropes across an important street, just making "the darkness visible" for those who ventured out.

Not that the people did not want light, especially in connection with their amusements, for night was naturally the time of leisure. Because of the dangers of night travel, as well as the inadequate interior lighting arrangements available, nothing much in the way of amusement could be arranged. Theatrical performances were generally given in the afternoon. It is difficult for us to realize the ingenuity that has gone into the problem of lighting the theater. In ancient times that misguided Roman emperor, Nero, took a demoniac pleasure in seeing people burned, and there are at least two paintings, considered authoritative, in which we find depicted the early Christian martyrs being burned as living torches to illuminate a circus in the Colosseum. That the temper of the time could be cruel enough to tolerate a deliberate brutality of this magnitude, that there could even be festivals if they needed to be illumined by the agony of living lamps, is hard to believe. And yet, does this not perhaps tell us, however poignantly, of the absence of any adequate methods of night lighting? Fortunately, while we want drama, and perhaps melodrama in the theater, this never became a popular course!

Because of the absence of light, the very logical procedure was to give performances out of doors, or at least in semi-open places. They were generally held in the afternoon, but always in daylight, for other than the dim, religious light obtainable from the use of candles, rush lights, or torches, there was nothing to serve the purpose, and these were useless for large-scale illumination.

In Shakespeare's time, when night fell, pitch blackness fell with it, with never a street lamp to shed a ray upon the road. People who had to go about at night had to take their own torches or lanterns with them. Therefore, in a play, although acted in broad daylight, a character would bring on a lighted torch, merely to indicate to the audience that the scene was supposed to take place at night.

At the time of our great-grandmothers, the theater knew neither gas nor electricity. Before the chandelier with its patent oil lamps came into vogue, a "barrel hoop" suspended over the audience served to light up the evening gowns and reflect the jewels of the assembly. The crude lamps used at the wings of the theater had a dangerous way of flaring out and threatening to burn up the make-believe palaces, trees, and sky. These lamps were open "floats" with wick-holders coming from the bottom, and the oil, when heated, would only too easily catch fire. To guard against

The Graphic (1881)

The Savoy Theatre, London, the first theater lit by electricity, during a performance of *Patience* by Gilbert and Sullivan. The theater was opened October 10, 1881; **note** the pear-shaped globes along the upper balcony.

accidents, a tub of water was always placed on either side of the stage, with a large mop ready for use at a moment's notice, and it was rarely that a night passed without its being called into use.

An extraordinary number of fires occurred in theaters about this period, and a list of them was published showing a loss of life of more than nine thousand people in a relatively short time. Since then, particular care has been directed to safeguarding the theater, and of course electric lighting lends its constructive aid in this endeavor. The old system of a row of "footlights" with their unpleasant upward shadow is now obsolete. In those days of lamplighting the actors were obliged to stand in a line in an effort to get into the small stream of light shed from the wings and assisted by the "float" footlights, which they called the focus. Gas-lighting was introduced in a Parisian theater in 1822; then came the limelight for special effects. Although a gas-lighted theater nowadays would seem to us primitive and entirely inadequate for modern productions, it made a great sensation when first known. The Old Bowery Theater was the first amusement house in New York to be lighted with gas. This was in 1825. Electricity was employed in 1882 in the Savoy Theater, London.

Perhaps to the theater can be attributed our present symbolic interpretations of the qualities of light. The producers of plays in Italy at the time of the Renaissance recognized the symbolism of lighting, and knew that many of their effects must depend upon the judicious placing of lamps or candles. They illuminated fully those scenes in a play in which the subject matter was happy; and when in a tragedy the first unhappy situation occurred, the lights were dimmed to induce melancholy or at least a pensive and serious state of mind in the beholder. Ever since then, lighting has been appreciated in its artistic and psychological aspects, and today effects have reached great beauty and significance.

Theater lighting today uses all the inventions in the field of optics. The colors are chosen for their effect by psychologists and are harmonized with the mood of the music and the action. The keyboard of switches, much like the keyboard of an organ, is in the charge of an expert electrician who must know the producer's business as well as his own. Cues for light changes must be rehearsed, and care taken to have just the right intensity on playing areas. Modern stage-lighting, greatly influenced by Norman Bel Geddes, is a complex science and a highly developed art which can contribute greatly to the success of a production.

More and more is light assuming both decorative and psychological importance. That it has great power to affect one's moods needs hardly to be demonstrated. Moonlight, firelight, rosy dawn, the veiled light of a rainy day, the steady, low light of a museum, the bright lights of a great city, all affect us differently, evoking various and characteristic sentiments. Just as we would grow obtuse to any one *natural* light without change or variation, so, too, artificial light may now be largely arranged to suit one's varying moods. Travelers who visit the land of the midnight sun tell us that the Scandinavians have a serious, heavy attitude toward life because of the monotony of unending nights and unchanging days. Climate we cannot change, but with the flexibility and efficiency of modern lighting we may nowadays obtain in every room at least two different lighting effects in order that unlike moods and circumstances may have their proper expression. Lighting is no longer merely an expedient. The hard struggles to penetrate the darkness are over. Today we give ourselves to those broader applications of it that yield mental comfort as well as physical—the restfulness and peace that come with safety and protection.

Knowledge of what light is, and how it may best serve our purposes, is increasing. But we continue to look to it as a symbol. The first order that was given was "Let there be light." How far we could go on that one exhortation! Seeing always beyond the mere word of command into the spirit of it, let there be also light that shall dispel the darkness of ignorance and intolerance: torchbearers all of us, lighting the way, and illumined too by the flames of those who go before.

V

Milord and Lady Dine

THERE WAS ingenuity in the traveling salesman who, as the dinner guest at a buyer's home, was asked to say grace. Being unaccustomed to the practice of this ceremony, he could but do his best. "Dear Lord," he began, "we are in receipt of your kind favors of recent date and beg to thank you. We hope to merit your continued courtesy." When we survey the story of table manners, we see that those who have gone before also did their best, even when that was not all that it might be.

Without beginning in that remote antiquity when man would go forth to catch his dinner and then hide himself while consuming it in fear and superstition, there are countless stories to be told of more recent times, showing how mother wit has contributed to the romance of table manners and of culinary curiosities.

In ancient Greece, and through her later in Rome, it was the fashion not to sit at table as we do but to recline on couches. This came about when the Greeks would go to their great public baths and thence at once to rest upon a couch. While resting, a light repast would be served them, and this easy attitude was thus introduced into the home.

The dining table increased its prestige in the Middle Ages. In medieval times, when there were no hotels in the present sense, a traveler had to take his chances at any inn or tavern that the town might possess. He was usually welcomed at the table of the innkeeper or host himself, and joined informally in the conversation of the family and guests, accepting such "pot luck" as might be provided.

Table d'hôte, that familiar expression today, meant the table of the host. Ordering *à la carte*, or by the card, at a time when the guest was undoubtedly unable to read, would have been rather like inviting the fox to dine out of the stork's deep vase.

It was not until the sixteenth century that the custom of providing a written or printed menu originated. At a meeting of the Reichstag in Germany at that time someone noticed at a banquet that the host, Duke Heinrich von Braunschweig, had before him a slip of paper which he glanced at now and then. On being questioned, he replied that it was a list of the dishes that were to be served, made for him by the chef so that he might save his appetite for those he liked best.

At the time of which we were speaking, however, travel was so beset with difficulties, and culinary art so lacking in variety, that the idea of choice was nonexistent. A person was glad of anything he could get. At that time a wayfarer was by no means even certain to find an inn, and it became a convention that hospitality was generously accorded everywhere. It centered around the *table d'hôte* of the feudal lords. Every householder, particularly the well-to-do landlord, provided accommodations not only for his own retinue but for strangers who might at any time knock at his gate. In the great medieval halls, for about a quarter of their length, the floor was raised by a step. This space or dais was occupied only by the principal members of the family and important visitors. The great table that stayed permanently here at the upper end of the hall was covered with a scarlet cloth and placed transversely across the platform. At right angles from its center ran the longer and lower board at which domestics and inferiors fed. The arrangement resembled the letter T in form.

As long as the great hall was the chief and often the only room in the house, and therefore the central point of household life, it was not maintained in its dining aspect all day. It was desirable that the furnishings should be easily movable—a feeling we can sympathize with in this day of combination rooms—and this explains the fact that the table of those times consisted of a long board placed on trestles that could be easily taken away and stacked at one side.

A conspicuous peculiarity of medieval dining was the fact that the guests sat only on one side of the table, the other side being used for the service. When there was a great banquet the tables were often arranged in the form of a horseshoe, both to facilitate the service and to provide space for the dramatic performances or "cabarets," as we would call them today, which were often introduced between courses. There are records of round tables, but these are not our modern conception. They were simply narrow tables arranged in a circle instead of a horseshoe, the inside being left with openings for attendants to serve the guests who sat on the outer edge. This was the kind of round table about which King Arthur held his court and entertained his famous knights.

The host and guests sat with their backs to the wall, facing the entrance, as a measure of protection against an entering enemy. We can imagine the "festive board," literally a board, its guests strong fighting men, their armor piled on the floor near by, gay banners hung upon the walls, the dogs picking up the bones and crumbs, and loud music or, more likely, "joyful noises" contributed by some of the domestics. And until the fifteenth century, and later, it was a rule of etiquette to dine noisily: "smack thy lips resoundingly if thou wouldst show due appreciation to thine host."

Today a meal is automatically accompanied by knives, forks and spoons. Yet with what scant attention do we regard these implements.

In olden times the most sumptuous repasts were served without cutlery, and until even the middle of the 1600's the eating habits of all classes of people, including kings and queens, were quite primitive. Dining, in contradistinction to eating, is a rather recent art. A knife blade and a spoon were considered equipment enough, and these were freely assisted by the fingers. Large basins, ancestor of the finger bowl, were passed around during the meal, together with napkins that several people used in common. Of chief importance was the spoon, but, curiously, no hostess felt obliged to supply one to the persons present. On the contrary, the only way to have a spoon was to bring one yourself, and the kind you brought was largely instrumental in deciding your place at table. A silver spoon was the equivalent of an admission ticket to the aristocratic section of the table. Most spoons were of bone, shell, or other materials, so we can see that the expression "born with a silver spoon in one's mouth" really meant something, for gifts of silver spoons were wonderful birthday presents and often changed the social status of the recipient.

Where did the first spoons come from?

People of very early times had congregated along seacoasts and rivers. These waters provided shells which served as natural spoons with which to eat certain types of foods, and were available long before man knew how to fashion metals into needed tools. To make shells more convenient, wooden or bone handles were doubtless attached, which produced an implement very much like the spoon of today. At various times metal spoons have been designed reminiscent of this tradition, with a shell depicted in the bowl of the spoon. In the forest lands, scooped-out wood was used. All early spoons were larger—more like the modern ladle.

Knives were first fashioned of flint and other suitable stone materials, and, as knowledge progressed, of bronze, then iron, and later steel. It is not definitely known whether some of the early knives were put into practical use by the owners when eating. It seems reasonable to suppose that they may have been, but there is plenty of evidence that the knife was not until the Middle Ages regarded chiefly as an eating tool. It was doubtless used to cut meat or other foods prior to serving them, but we are told that they were not part of the individual's eating equipment. Food was roughly torn apart with the fingers and teeth—a pretty picture to contemplate. The Chinese, more delicately, use chopsticks because they consider the knife and fork barbaric. "We sit at table to eat, not to cut up carcasses," they say. When knives were finally employed as an eating tool, the same knife which was carried in the belt as a weapon was the one that was used for this purpose. Knives to be used as weapons ended in a sharp point. This point served instead of the still-unknown fork to spear the food and hold it steady, or to carry it to the mouth.

The introduction of the fork can be more definitely traced, since its history is not so venerable. Large forks have been found in the remains of early civilization, but these were used as weapons and not for eating. The fork as a weapon is well known to everyone because of its association with such mythological characters as Neptune, Triton, and the devil. Large forks were known for raking hay or for taking meat out of pots.

The fork is also known to have been used as early as A.D. 600 in connection with viands, but it was regarded for a long time only as a useful utensil for serving them. At least, that is true of forks made of metal. It seems logical to suppose that man used a one- or two-pronged fork in the form of a stick to hold food over a fire, and no doubt food was often eaten right from the stick. But ingenuity did not dictate its use in small size as a table utensil until about the sixteenth century. At that time it suddenly appeared in Italy, introduced probably from the Orient among a few members of the nobility and upper classes. It had been used for some time before that, in rare instances, for the eating of special dishes and delicacies such as preserved pears.

The first forks were made with only two prongs, like the carving fork of today, and it is no wonder that our ancestors had to eat peas with a knife. Up to the end of the sixteenth century, forks were a novelty and a luxury, and were considered in the light of amusing toys or curiosities rather than as creations that might have utility. If used at all, it was to aid in cutting, not for eating. At first, forks were very rare and hard to come by. A collector might have sixty-nine knives but only three forks. Around 1423, a Spanish nobleman announced the discovery that these instruments would enable one to carry cooked meat to the mouth without greasing the hands, and he seems to have been the first person to be struck by this happy thought. But forks were not taken up by the common people at first, because they were regarded as effeminate. Even the pulpit declared against forks, stating that their use was insulting to God, inasmuch as they substituted mechanical means for the heaven-given boon of fingers. But progress was not to be downed, and they finally pried a way for themselves.

It can hardly be realized to what extent the fork has influenced manners and the culinary arts. So long as only a knife was used, it was customary for the person carving to place a covering of paper on the part of a joint which was seized by the left hand. This was not difficult in the case of a leg of mutton, for instance, which offered something to grasp hold of, but in the case of beef it was not so simple. The paper frills which we still see served on chops are a survival of this custom of the carver to protect himself. After the innovation of forks this was changed. Carving became a finer and more elegant art, and the napkins little more than a decorative accessory. It became bad manners to put the fingers to the meat dish, and worse to carry meat from the plate with these implements of God. With the coming of the fork it became possible to enjoy a taste for solid foods. Previously they had to be cooked to a soft stew so as to be manipulated with a knife and spoon. It was now easy to have delicate slices instead of clumsy lumps, so that the natural flavors of the foods might be discerned and appreciated.

The fork of course promoted cleanliness unknown in the days when the fingers were the only dining equipment, and incidentally made it possible to use finer table linens, handsome napkins, and choicer cloths. If we still had to be forkless today, our formal dinners would lack all the decorum and charm that make them the gracious, dignified events that they have become. Nothing has had a greater influence in transforming the biological necessity of eating into the cultural, social art of dining than the development of the now commonplace knife, fork, and spoon.

Even plates were by no means in common use. Individual plates were practically entirely lacking until the sixteenth century. It was customary for two persons to share a *couvert*, which consisted of a thick slab of bread cut from a large loaf, on which the meat was served sizzling from the fire, and which was later cut in halves and consumed. The French word for a slice of bread on which meat was served was *tranche*, and from this was derived "trencher." This was a platter of wood, often with two hollows so that meat could be served on one side, and later pudding on the other, thus forming the origin of the "blue plate" of today.

Then there was the matter of salt, which has been humorously defined as that which, if not put in the soup, spoils it. Besides the general utility of that indispensable condiment, and apart from scruples with regard to spilling it, salt was regarded with profoundly superstitious feelings; it was considered desirable that it should be the first article placed on the table after the cloth was laid. The ancients spoke of it as divine. Our word "salary" is literally salt-money. The salt cellar made for Francis I by Benvenuto Cellini is one of the most outstanding of his works. To sit at table above or below the salt was a social distinction that was quite definitive. The salt was contained in a massive dish caller a "saler," later corrupted into "cellar." Important guests sat between it and the host, "above the salt," which was placed centrally on the table, while servants and socially unimportant folk sat below. This was more for decorative and symbolic value than for use, as smaller "trencher" salts were placed within reach of the guests for practical purposes.

About this time there was a pleasant custom of poisoning persons in power whom you did not care for. As a result of this, many high personages caused the eatables and beverages that were going to be served to them to be tasted first in their presence. This testing, which enabled them to believe in the safety of proceeding, took place on a special piece of shelved furniture called a *credenza*, or credence table. "Credence" comes from the Latin word *credo*, which means "I believe" or "I trust." It is the same word we use when we give someone credit. Hospitality today requires host or hostess to partake first of wine or food for similar assurance. (Unfortunately, today's additives do not work so fast nor so dramatically, but are more insidious.)

In lordly or rich dwellings, credence tables were utilized not only to test food but also to display the silver and table services, and thus became a sign of wealth. A recognized rule seems to have prevailed: so many shelves were allowable for counts, so many for dukes, so many for princes, so that a "social register" or "blue book" wasn't really necessary!

While these customs were establishing themselves, there came a gradual increase in the variety of things available as food. Until the Tudor period, few vegetables were cultivated in England. Then Henry VIII sent to Holland for gardeners and started the cultivation of vegetables a little more generally. Again, we seldom realize when consuming our morning coffee or afternoon tea (if any) that these cheering drinks were, with cocoa, introduced into England within a few years of one another about the time of Charles II. Until then and for some time afterward, for they were at first too luxurious for staples, breakfast consisted of cakes and ale—rather cold fare for a foggy morning.

The introduction of coffee was responsible for a new kind of social life in England. Coffeehouses were opened all over seventeenth-century London. The first coffeehouse was set up by the servant of a merchant trading with Turkey. He had accompanied his master home from Smyrna to prepare his coffee for him in the morning. This excited so much public attention that the servant was allowed to open a coffeehouse, and the example was freely followed. Coffee-

N.Y. Public Library

Coffeehouse interior (17th century)

Humphrey and Vera Joel

A caricature of Lloyd's Coffeehouse in 1798.

houses became, so to speak, clubhouses free to all who could buy a cup, and yet each was known for its special circle of visitors, literary, scientific, religious, or political. In the absence of newspapers, they were a great means of spreading news and discussing public questions. Many from the middle and higher classes attended them daily, and they came to exert a powerful influence on public thought.

The coffeehouse reached the zenith of its popularity in Queen Anne's reign. It was the center of news, the lounge of the idler, the rendezvous for appointments. Men might have their mail left there, and might rub shoulders with almost anyone. Here they were considered equal and took the first seat that came to hand. If a man swore, he was fined a shilling. Quarrels were prohibited. No card-playing or dicing was allowed, and no wager might exceed five shillings.

Of even greater permanence was the influence of tea about the same time. It is at first glance difficult to see how so trifling a thing as a cup of tea could have influenced decorative styles throughout a continent for the greater part of a century, and yet this, oddly enough, was one of the results of the importation of tea from China. For up to that time very coarse pottery had been in use, but the popularity of tea brought it into too much prominence. Someone had the idea of importing some Chinese porcelain to drink the tea from. Only to see that exquisite work was to want it. A craze of china-collecting began. Having the porcelain, where to put it became the question. Why, of course, the Chinese lacquer cabinets! Then the rugs and other accessories logically followed. It became smart to have everything decorated in *chinoiserie*, the Chinese taste. Pagodas and mandarins appeared everywhere. Chippendale created the style we call Chinese Chippendale, which carried the tendency still further. Those who could not afford the originals of all these things tried to make imitations. All the china made in Dresden, Sèvres, Chelsea, and elsewhere in the eighteenth century resulted from this effort. The influence swept gradually through every stratum of the decorative arts. And all from the little cup of tea!

Sugar, of which the yearly average for each person in America was once estimated at the surprising figure of 110 pounds, has not known common use for much more than a couple of centuries. That, according to health teachers, is about 110 pounds too much, as this pleasant substance is bad for the teeth, bad for digestion, bad for the complexion, and bad for the figure, and even the energy it is alleged to provide is illusory. It was used by the ancients chiefly for medicinal value, honey being used in those times to answer the craving for something sweet.

Had it not been for the camel in Arabia, the world might be without another product that is commonly used today by those who are not cholesterol-conscious. For the origin of butter has been traced to the Arabian habit of carrying sour milk in skin bags across the desert. The swaying motion of the camels caused the contents of the bags to turn to butter, thus giving us a product without which many feel that present-day cookery would be at a distinct disadvantage.

Modern culinary art was born in Italy at the time of the Renaissance, when wealth and ease turned the attention to luxuries of the palate. Venice was the door through which oriental luxuries entered Italy, but at the same time there were culinary traditions which the Italians had direct from their own ancestors. Sicilian cooks were favored by the ancient Romans, just as French chefs are in modern Europe today. Catherine de' Medici it was who brought its higher branches from Italy into France, where it fitted exactly with the French temperament. The motif was played upon and elaborated there to its highest development.

Many famous Frenchmen, including royalty in the persons of Louis XIII and Louis XV, Dumas, and Talleyrand, have demonstrated the fine art of dining, and gastronomy as an intellectual pursuit was launched upon its famous career in France in 1803 when M. Grimaud de La Reynière produced his *Almanach des Gourmands*, which was for many years a successful publication and did much to establish dining as a fine art in contradistinction to eating. In 1826 Brillat-Savarin published that brilliant work, *La Physiologie du Goût*, in which with felicitous skill he discourses on the pleasures of the table and dwells on its scientific aspects too, as of that period. He was, like La Reynière, a lawyer by profession, and for quite a while had been exiled for his political opinions. As a young man he had studied not only law but also medicine and chemistry. Music, the fair sex, and good dinners all were of more than usual interest to him, as many anecdotes testify.

To Brillat-Savarin eating was not merely a pleasure. Rather did he feel it a solemn and exquisite duty which one owed in appreciation to Nature for her bounties. It is evident in his writings that he was keenly eager to teach people the potentialities of the art of dining in its endless phases. His recipes show a touching anxiety lest some ingredient be misused or left out. To him the cook is a real scientist with an infinite capacity for effecting good or evil. Brillat-Savarin's style of writing is as excellent as his original researches were valuable, and proclaims him the genial companion and ready raconteur. His works stand not only among the most original and important of the nineteenth century, but, despite the greatly expanded literature of gastronomy, they are still one of the main sources of instruction for those engaged in any aspect of this subject.

Alexandre Dumas was one of the famous *bons vivants* of his day, and he was never so happy as when discovering a new recipe. His reputation as author of *The Count of Monte Cristo* and *The Three Musketeers* was not nearly so dear to him as his reputation as cook, host, and epicure. "I wish to conclude my literary work of five hundred volumes by a book on cookery," he told his friends. Later, rejoicing openly, he declared, "I see with pleasure that my culinary reputation is increasing and promises soon to efface my literary reputation." The last volume he ever wrote was, indeed, a cookbook filled with recipes gathered from all over the Continent.

Talleyrand, French diplomat and politician, was pleased to be called "the first fork" of his time. When he was eighty years old he was accustomed to spend nearly an hour every morning with his cook discussing the dishes that were to compose his dinner. He believed that a well-devised cuisine supervised by the best available chef would preserve his health and prevent illness better than a staff of doctors. For twelve years the famous Carème held the position of culinary director for him. Carème had been at different times chef to Czar Alexander I of Russia, and to King George IV of England.

Then there was Baron Brisse, a *gourmet* so well fed that, so the story goes, he was obliged to buy two seats in the diligence when he traveled. Henrion de Pensey, a magistrate of France, was of the opinion that "the discovery of a dish that sustains our appetite and prolongs our pleasures is a far more interesting event than the discovery of a new star, for we have always stars enough."

At the other extreme of the school of gastronomists came the fourth Earl of Sandwich, whose very name must have forced wrinkles of contempt in the noses of those we have been discussing. This notorious lord had but one passion, the gaming table, and devised for his convenience a tasty combination of meat and thick slices of bread in order to avoid the interruption for dinner.

From the Russians the Paris restaurants of today borrow the custom of beginning a meal with hors d'oeuvres, or appetizers. Cold smoked salmon, tunny fish, sardines, baby artichokes in oil, various vegetables, fish and lobster salads, cold eggs, sliced sausages, and sundry other delicacies comprise this feature. While all these things may be delicious, it seems a misnomer to call them appetizers, for in such quantities they are also appetite destroyers, especially if prefaced to a substantial dinner. In Russia and Scandinavia, where extreme cold creates a ravenous appetite and a great capacity for stowing things away, it is undoubtedly a good custom; but the transfer to a French or Italian table is an artificial one, whereas climate would be a safer guide.

An essential difference between French and English cooking is that an English cook is naturally economical in time and trouble, and a French cook in material. Once a chef and a maître d'hôtel were having an argument. Said the chef: "But, monsieur, this fish, it is no good. I cannot serve it for the dinner." Said the maître d'hôtel: "Of course you will serve it: any chef can make a good fish taste well, but the skillful chef is the one who can make a bad fish taste well."

In the matter of breakfast, England has conferred a blessing. The Continental habit of eating only bread and coffee for breakfast and then having a second, heavier breakfast during the morning has always seemed a little queer to me. Normal people have a good appetite in the morning after a night's rest. They should be able to gratify it and go peacefully about their business without further interruption until lunchtime. This custom we have acquired from England. When you travel in Italy or most other places on the Continent you always wonder whether breakfast will consist of rolls and honey, or honey and rolls. This seems strange to Americans, to whom the modern breakfast is something quite different. The rolls and honey are almost as unsatisfactory as the cakes and ale of old England would have been.

An important factor in making modern breakfasts interesting if not healthful is the use of so-called breakfast foods, the cereals so diversely prepared. Where once grains were served only as mush or porridge, today they are puffed, rolled, cracked, ground, shredded, malted, or flaked. So variety, always a valuable feature, is added to or taken from the actual nutriment, which does, however, add to the cost.

But a great surprise is in store for you in the matter of

breakfasts when you come to Holland. A Dutch breakfast is like no other meal in the world. A long special table is set aside for it in a special room in a typical Dutch hotel, and it is decked with enough nourishment for a starved Roman legion. There are, to start with, about seven different kinds of bread: plain white bread, gingerbread, plain buns, buns with raisins, brown bread, toasted biscuits, sweet cakes, and a kind of fruitcake. You will find at least three or four kinds of meat—cold sliced veal, roast beef, ham, and sausage. Then, if you don't see it all at once, you smell the cheeses, two or three of the Dutch specialties, including Limburger and Edam. There are several dishes piled with boiled eggs, and of course there are jams and honey and seasonable fruits. The climax is a pot of steaming coffee, not the insipid beverage of the rest of Europe, but the real thing from the Dutch colonies.

To a Dutchman all this is trivial. It is merely a "breakfast" to stay him till a (to us) mighty lunch, and this in turn is something of an appetizer toward the grand climax of the day, which occurs at dinner. It would seem that the *huisvrou* counts neither cost nor calories.

One cannot help reflecting in this connection that many, not only in Holland but right here in America, are victims of painful diseases resulting from gluttony and overeating. They would be willing to give large sums of money to get rid of the disease, but they refuse to curb their gluttonous instincts. They appear to believe that they can gratify an abnormal taste for rich and unnatural foods and also maintain health. But they defeat their wavering purpose by allowing thoughts and desires which inevitably conflict with it.

In Germany, if you take any of the delightful motorbus trips about the countryside, you will find, as I did, that it is customary to stop three times a day for the sole purpose of eating, and each time a long-drawn-out event is made of it. Many of these bus excursions begin at 7:30 A.M., which is rather an early hour to be down at the starting place and ready for a day's sight-seeing. At ten we would stop for beer or coffee, but instead of consuming it in fifteen or twenty minutes, which would have been ample time, the Bavarians linger and loiter over it in their leisure-loving way, and remain for a full hour. At noon the stop for luncheon is nearly two hours, and at four o'clock there is another hour's pause for coffee. Altogether, in the average twelve-hour excursion, four are dropped thus, and I for one thought it would have been more reasonable to leave at 9 A.M., and return at six, eliminating some of these repasts. But I did not find that the eating itself was so important as the opportunity it affords for long sociable chats and conversation. The Germans do not eat a great deal, but they make it last long, and seem to enjoy the ceremony as the feature rather than the food *per se*, though this is usually excellent.

Whenever I travel abroad I am always impressed with the leisurely manner in which both social and business activities are conducted in foreign lands. Nothing seems reason for haste or speed. Even money-making itself is subordinated to convenience, and business is completely suspended at noontime and teatime and these other recesses. Try to buy a bill of goods from an Englishman at four o'clock. You do not exist for him at teatime.

A few apostles have been preaching that we get less out of our lives than the more serene folk of England and these other nations. We live too hastily to live at all. Leisure is a great thing. During relaxation we can gather ourselves together for continued effort. If we organize our time to get it, and then use it properly, we will find that during leisure we are laying plans for more system and method and greater efficiency for our periods of activity. It might be well if we could enjoy such recesses as do the Europeans. Our coffee breaks are not quite the same. After all, the purpose of business is to make a living, even if we do often forget the end by overstressing the means. Let us not forget to live while we are making the living. Let us, as we can, appropriate the virtues of these well-tried European customs, using our ingenuity to adapt them to our own needs.

In the process of adapting things, many cities have done this so effectively that they have come to be identified with specialties created by taking advantage of native products. Our popular hamburgers, of course, came to us originally from the German city of Hamburg.

Such delicacies as *Marzipan* in Lübeck, *pâté de foie gras* in Strasbourg, *Lebkuchen* in Nuremberg, and countless other things make many of these wonderful old cities even more famous than they would otherwise be. The *Lebkuchen*, or little spice cookies of Nuremberg, are a memorable delight. They are sold in the most picturesque tin boxes depicting the early grandeur of that fine old city, and the contents of the boxes are more than good, each box containing an assortment of the little cakes, one better than the next. Among other great German specialties are *Würstchen*, the famous German sausages. They are sold everywhere, even in the railroad stations, often with beer and pretzels.

The ancestor of our "hot dog" is seen here in all his glory. There are probably a hundred different kinds of *Würstchen*, obtained by using different meats, variously blended, spiced, and cured. Nearly every locality has its favorite. A German delicatessen store must be an epicure's delight. Officers of the health boards visit the delicatessen and food supply stores of Germany at unexpected times, taking away samples for chemical analysis. Fines are inflicted for the least lack of obedience to the pure-food laws and one may have perfect confidence in everything displayed. The immaculate cleanliness of the stores themselves is conspicuously apparent.

The Italians are also holding their own among the chief gastronomic nations. In London, Paris, and New York, as well as many other large cities, Italian restaurants are well patronized by non-Italians as well. One of the Italian specialties is *minestrone*, a thick soup made of all sorts of vegetables. This is a chief article of diet among the peasants of Italy, where the vegetables are readily accessible. In Naples one can at any time see the macaroni thrown over reed poles on which it is exposed to sunlight to dry. Macaroni is of course only one of some dozen varieties of *pasta*, including spaghetti, vermicelli, ravioli, and many others.

I have often wondered why an edible so commonplace as spaghetti should be served in the long strips that make its intake into the system so much a matter of juggling, fishing, or other sportsmanship. It would seem that common efficiency would demand its being served in more negotiable lengths. But we must remember that speed and "getting through," quick lunches, and the like are not characteristic of the European. The approved Italian way of eating it is gently to spear a spoonful with the fork, convey the end to the mouth and suck it in without much waste of time, biting it off at a convenient length and letting the surplus drop back to the plate. Which is expedient enough if not the last word in elegance. Another way is to wind it round your fork until there is just enough to fit into your mouth.

But there is no loss of flavor if the spaghetti is cut into convenient pieces and eaten any way you please. It is the usual noontime lunch in Italy, served in a countless variety of ways, and contains a fund of sustaining powers that admirably suit it for a prominent place on the menu, particularly if you are not having to consider your weight. An Italian *gourmand* who was once remonstrated with at dinner for eating macaroni after a hearty meal replied, "Madame, macaroni is heavy, it is true, but it is like the Doge of Venice: when he arrives one must make room for him."

The Italians, on the whole, eat meat rather moderately, as befits their warm climate. They are able to buy it in very small quantities, using just enough to impart the flavor of veal, lamb, or chicken to a dish of vegetables. Instead of meat, however, cheese is used freely as an ingredient of various cooked dishes, as well as dessert in the French way. And we know now that cheese is twice as nourishing and sustaining, weight for weight, as porterhouse steak.

The Italian salads are very good indeed, largely because

of their fine olive oil. They say that three people are necessary to mix a salad dressing: a prodigal for oil, a judge for salt, and a miser for vinegar. The Spanish add to this that the dressing should be mixed vigorously by a furious woman or a maniac. In Italy, although we had the dressings of olive oil, we had no olives. These, it appears, are saved for export, though the gnarled and ancient-looking trees added their silver-gray to every Italian landscape. But always there were oranges, and they were generally served picturesquely on their twigs, giving the touch of charm that one can get in the midst of abundance when the fruit need not be boxed and crated over long distances, but may be plucked as you go. And I suppose everybody knows the delights of that supreme Italian ice cream known as *spumoni*. It is indeed the perfect climax to a good dinner. Of course strong-willed or health-conscious folk will not indulge in most of the aforementioned specialties, but they do seem part of the record.

Today the niceties of dining are giving way to TV dinners—frozen combinations in aluminum foil, which have reduced a meal to the minimum of care and time in preparation. I suppose a leisurely meal with all the due amenities would not harmonize too well with the blatant type of many TV programs. As for letting "good digestion wait on appetite," digestion nowadays waits for nothing. The hasty breakfast, the rapid coffee break, the quick lunch, the TV dinner, the reducing and starvation diets bolstered with all the wrong things for continuous snacks, have made a mockery of dining.

As for real dining, this is largely taken out of the home in any formal sense and is done in restaurants and hotels. Such persons as Ritz, Statler, and Hilton have been among those who have glamorized dining and other hospitality services 'way beyond the plains of normal restraint, up the mountains of abandon, leaving Puritanism far behind.

It is gratifying that Puritan ideas no longer give us a sense of guilt when we enjoy the blessings of the table.

The father of Walter Scott, we may remember, was such a Puritan. When young Walter one day expressed his enjoyment of the soup, his father promptly mixed it with a pint of water, "to take the devil out of it." But we have learned to mix a bit of art with our vitamins, and add beauty and charm and understanding of the varied elements to the important function of dining.

VI

Outwitting the Lean Seasons

THE WORD "diet" as we understand it is a modern concept. Old-time people ate what they could find, and largely built their settlements where food was obtainable. Those who live where there are coconut trees need no other resources for a complete life according to their standards. The coconut tree has been called the most valuable tree in the world. It literally provides food, drink, shelter, clothing, candles for the native hut, oils for beauty or medicine, refreshing drink or hard liquor, rope, soap, toys for the savage child, and curios for adult pastime. Even the uncivilized man used his ingenuity to utilize every portion of this precious tree when other facilities were lacking.

But most communities lack such natural provisions. Wondering where the next "meal" would come from was a chronic state for primitive man. He was lucky if he happened to be born in a land of plenty, but those were few. There, if the weather was favorable and the fruit grew, and the animals held still, all was well for a time. But a drought could burn the vegetables and starve the animals, or a flood could drown out the land. And when winter came and the trees were bare, there was nothing but fear and hardship. He ate what he could find, and moved from place to place to find it.

From the most primitive times the need existed for preserving food over longer periods than nature provided for. Sailors of long ago halted on their voyages as the food supply drew low, to plant and grow a crop—a quick one that was ready in six weeks or so. During this time they had opportunity to learn about the land where they were; then they resumed the voyage. This was interesting, but because foods could not be preserved to take along, considerable time was lost.

As soon, however, as edibles were gathered from field or jungle, it became necessary to conserve them for use between crops or in times of scarcity. We think of the intricacies of modern canneries without realizing how acutely early man was confronted with this problem, for he was isolated and fairly helpless—undeveloped in experience and without precedent to work with.

But there was his ingenuity. The animals roved for their foods, and man was long obliged to follow them. Then the hunter, bringing his kill into the hut and hanging it in the smoky atmosphere of the hearth, found that the heat dried it, and the smoke itself did something to it that deferred deterioration. A purely accidental discovery, but one that his mother wit enabled him to build on. After this, less roving was necessary. The nonproductive seasons held less terror. Other similar discoveries came to his aid. Salt was found a preservative; milk could be kept in the form of cheese and butter; grape juice could be saved as wine.

But the facilities were not those that made for health. Such foods might stay the hunger, but they were so limited, so ignorantly combined, that no really constructive dietetic efficiency existed. If anything did not agree with him, he assumed that the gods were angry about something. This merely added a little extra fear to an already ample supply. Many must have been the pioneers who suffered martyrdom by trying a new "dish" which turned out fatal. The want of fresh vegetables and fresh meats undoubtedly had a large share in making the people susceptible to those scourges and plagues that attacked them from time to time throughout the Middle Ages.

As long as foods could be preserved only by drying, smoking, or salting, vegetable foods could only be had in limited quantity and unpalatable forms, while during most of the year the meat diet consisted of salted products, unappealing to taste, and hard. There was no way of transporting perishable foods without deterioration. It was standard practice for everything to taste bad even before much discrimination was developed. Even when reasonable variety did become available, little thought was given to the idea of diet in a modern sense. Foodstuffs that could tickle the palate were such a novelty and pleasure that nothing else was even thought of as a food virtue. Calories, vitamins, digestibility, nutrition, by these names or any other, bothered no one. But had there been movies then, the great motive would have been not a sex thrill but a taste thrill. Primarily it was the craving for seasonings and spices that caused explorers to voyage to the Orient, as the spices were absolutely necessary to make the tainted, monotonous foods barely palatable.

Our forefathers of the Middle Ages would be amazed to be set at a well-organized meal of today. Man's ingenuity has demonstrated the fuel value of food, the nourishing qualities of various edibles. Men are learning to eat for efficiency. Mothers are learning the relation of foods and feeding to the health and development of their children. Many interesting experiments in manufactured health foods are being made today. Those originating in the laboratories at Battle Creek cover a wide range and have established an unusual note in dietetics. Among other things, they have produced a number of foods designed as meat substitutes. The matter of meat-eating is the subject for much scientific inquiry at the present, and there are widely conflicting evidences being sifted down. Stefansson, the Arctic explorer, tells us his adventures with meat-eating. Instead of taking six tons of provisions along, he took six pounds of shot, so to speak, and gathered what he needed. But he had to be satisfied with whales and polar bears, and in the whole time he was away he had no vegetables at all. On his return this proved of interest to physicians, who tested him and found him entirely unaffected by this strange diet. On the other hand, spiritual people in some lands abstain entirely from meat, like the Hindu yogi, and achieve rather remarkable results in physical control, mental clarity, and spiritual unfoldment.

Nowadays the reducing diet is perhaps the most popular (or unpopular). Many people, however, do not mind how fat they may be, so long as they do not *look* fat. For these the costumer rather than the dietitian is the rescuer. But diet is more seriously considered today than ever before. Our babies are fed by all the vitamins in the alphabet, and all through life we may, if we are interested, eat the thing to effect specific purposes of health, strength, size, beauty, intellect, or what you will.

A very helpful factor, as well as an interesting example of ingenuity, in the business of dining is the matter of preserving and canning foods in the modern way. The object of farming is to create a dependable surplus of food. Ingenuity has devised constantly improving appliances to do this. The digging stick, the hoe, the plow, the mower, the reaper, the harvester—all have been developed in response to this need. Methods, too, are created toward the same purpose. Fertilization of soil, improvement of grains, rotation of crops, are some of these. But many of the farm products are perishable, as are also meat products. Only by taking measures to preserve them can they be advantageously distributed.

With the aid of even primitive preserving, early man could support himself through treacherous weather in latitudes otherwise inhospitable to life. Thus there is a decided bearing on the development of race and nation traceable to this ingenious procedure.

The methods of heating and drying, smoking and salting, were all that were at the disposal of early man, and until 1795 nothing was added to these. Then, however, François Appert, a Frenchman, undertook to preserve food in all of its original goodness without any outside elements. He placed various kinds of food in glass jars and bottles which he tightly sealed and corked. These he put in water which he heated for different periods. After the bottles were removed and cooled and kept for a long time, he opened them and tasted the contents. They were as good and tempting as originally. Since then, working on this basic principle, great industries have developed, and great services performed for the art of dining. Let us consider a few of these services and see how ingenuity, once started on the way to create, goes forward developing the constructive impulse.

The early problem of making plentiful harvests to last through the lean seasons is solved by the canner. Without him, thousands upon thousands of tons of provisions would be wasted because they could not be consumed when fresh, and later when they were really needed they would be scarce and costly. Meanwhile, the farmers would be able to produce and earn only at limited seasons, whereas now a continuous supply of produce can be disposed of to the canners.

They in turn are able to serve people in many lands that have not the facilities for satisfying their own needs. If only perishables were available, all this commerce would have to be eliminated, thus injuring the agricultural industry, depriving the recipients of the food, wasting the crops, and causing loss all along the line of procedure.

Many canners have branches right at the farms and can take care of a fruit or vegetable crop as fast as it is harvested. Such plants can put up from 250,000 to 300,000 cans a day. The men who grow this food are always sure of a market for their products, and people in remote industrial sections of the country where food is not grown can depend upon securing good, wholesome things to eat at moderate cost, without trouble, personal effort, or uncertainty. By the intervention of the canners they may have fruits, vegetables, milk, meats, fish, and other edibles in their homes in a state as pure, wholesome, and appetizing as when they were fresh from their various sources.

Whether people live in the desert or at the poles, they

may have their choice of it none the less. In bitterest winter the family in the heart of a Norwegian blizzard may enjoy pineapple raised in the distant tropics of Hawaii. And the canned pineapple, like other fruits, will be of better flavor than the fresh that we can buy at the market. These were picked green so that they can stand the journey and ripen to the proper point on the way. But the canned ones are picked ripe and preserved in their perfect condition.

Not only canning but condensing, dehydrating, freezing and other processes are also in use; but since the underlying idea and purpose are similar for all these, we will not go into the special virtues of the various methods. All such foods have been of particular use to soldiers, explorers, and others who want small baggage, safe transportation, and concentrated results.

It is not always realized how far-reaching an idea may be. Doubtless Appert foresaw nothing like these many ramifications of the work that he initiated. But it is one of the ways of success to ally one's self with the helpful and constructive thing, however apparently small. Since movement is a cosmic law, the very momentum which the idea gathers brings it toward success. This is of course only a preliminary factor. We shall indicate elsewhere (see Chapter XIV) some of the other prerequisites.

Ingenuity has contributed largely to the safe transportation of perishable foods that are not canned. Most foods have peculiarities which must be taken into consideration when long distances are to be covered. They are sensitive to either cold, heat, moisture, or some other condition of climate. To overcome this, special cars have been devised for the more sensitive products. Some are provided with refrigerating equipment, some with special ventilating arrangements, and the like.

As typical of the solicitude with which this is undertaken, consider the journey of the banana from its native Costa Rica. From this place some fifteen million bunches of bananas are annually exported all over the world. The fruit is cut according to the length and character of the journey. It will not do for any bananas in a shipment to become overripe before arrival, for hundreds of bunches can be spoiled by one bad one. A single banana ship carries 50,000 bunches in its hold. They are loaded by means of mechanical conveyors and are received into rooms already cooled to the right temperature by a system of refrigeration pipes and large centrifugal fans. This system keeps the cargo both cool and dry. When the fruit steamer reaches its port, the bananas are quickly transferred to cars properly refrigerated, or perhaps to cold-storage rooms in a great fruit warehouse.

A man is sent with every banana train to see that the bananas are kept in the best possible condition until delivered into the hands of the customers of the fruit company. These "banana messengers" must inspect the cars at division points, record the temperature, and take the proper measures to cope with weather conditions in the territory through which the cars must travel. These cars are made so that they may be heated as quickly as cooled. No *touriste de luxe* ever had more perfect attention than the personally conducted banana.

As for ingenuity, research in food has taken on an air of true magic at the fantastic farm maintained by the Department of Agriculture in Beltsville, Maryland. New types of birds, beasts, and bugs are designed to combine virtues and eliminate faults. If you want a turkey with all white meat, or bees with a gentle disposition, if you want to peel onions without tears, or have peaches without fuzz, or tomatoes immune to blight, these are only a few of the simpler achievements which are all in the day's work to these reformers of nature.

One of the outstanding examples of ingenuity with reference to foods is the use of by-products in connection with the preparation of meat in the stockyards. It is said that everything of a pig is used except its grunt.

The study of waste has been undertaken by skilled scientists laboring in well-equipped laboratories. Their research has covered every phase of the subject, and here again valuable industries have been created as a direct result. Since waste has been so thoroughly reclaimed, producers obtain more for their livestock, and consumers pay less for their meat. Among the valuable nonedible by-products of the packing house are such diverse things as fertilizer, soap, glue, hair for upholstering, bristles, and brushes, strings for musical instruments, hides and leathers, artificial teeth, buttons, candles, and wool. Some of the by-products of meat animals are exceedingly valuable, especially pharmaceutical preparations. Suprarenalin, a powerful astringent and heart stimulant made from the glands of sheep, is worth more than $5,000 a pound, but it requires the glands from 125,000 sheep to make one pound of it. The development of this by-product would be out of the question except under the large-scale production that is possible to the modern stockyard.

We think of the farm as an area where cattle graze, yield milk, hides, and meat, and on which edible plants are grown for our nourishment. Yet the trend is increasing that crops are cultivated fully as much to supply the raw materials of chemical industries. The versatility of corn extends far beyond its primary use as food and feed. More than thirty modern industries, from cotton goods to steel and explosives, now depend on refined cornstarch. Corn is used for many medical purposes from the manufacture of penicillin to sulfa tablets, diabetic foods, and so on. Cornstalks are suitable for paper and wallboard. Corn alcohol can be used in synthetic rubber and shatter-proof glass. We have named the merest few. So with wood, which is no longer merely for the construction of tables and chairs and houses. It is valued as cellulose which can be reconstructed into silklike filaments, cattle fodder, or explosives.

The ramifications of the influence of chemistry are incalculable. It is more than economic power, but enters into the customs and social life of our people. Where it can lead is a subject full of potentials for the future of our country and the world.

The transfer of the manufacture of food from the home into the factory, stockyard, and other public centers, and the need for transporting it over great distances, brought about at first a troublesome amount of adulteration. This made necessary government regulation to promote purity and truthful labeling. The very valuable work of Dr. Harvey Wiley and others in preventing food adulteration has conferred security and protection in food wholesomeness to such an extent that danger from this source is now negligible. But other sources of danger are still with us in the form of additives, insecticides, preservatives, synthetics, worn-out soils, and the like.

Yet information on the subject is growing enormously. Good nutritional practice, organic health foods, the place of vitamins and minerals in preventing and curing ills, and tenets of natural living are readily available to those who would put first things first. That great crusader for your health, J. I. Rodale, publishes the magazine *Prevention* in Emmaus, Pennsylvania, which everyone should read. Rodale is the man who brought organic farming to America; who, through his publications, tells us how to keep well rather than get well; who as a health researcher and crusader has told Americans hundreds of unusual and fascinating health secrets, such as the advantages of utilizing the sunshine stored in the sun-drenched sunflower. And in addition, he heads up a number of large industrial and manufacturing projects almost on the side. Through his magazines and books and public talks, Mr. Rodale has become the leader of an agricultural revolution which is of the utmost importance to the health of every one of us.

The government has recognized the work he has done in his Organic Experimental Farm. He has been called to testify in congressional hearings on various problems concerning food and agriculture. Visitors to his experimental farm have included former Secretary of Agriculture and

Vice-President Henry A. Wallace, Louis Bromfield of the famous Malabar Farm, and many business executives and congressmen as well as thousands of gardeners and farmers from all over the world.

A few "did you know's" may be given here to help you solve some health problems. These are quoted from his book *The Health Finder*, an encyclopedia of health information from the preventive point of view. It would take pages here, as it does in *The Health Finder*, to give the reasons for its well-documented advice. Anyone who wants the reasons can find them cogently expressed in that able volume.

It will surprise many readers to learn that oranges and grapefruit are not the desirable foods we think they are. They are particularly hard on teeth and gums. The juices are especially deleterious, causing erosion of the teeth and pyorrhea, as well as other bodily aches, itches, and some forms of dizziness. Without our entirely eliminating these pleasant fruits, they could be greatly limited, and vitamin C taken in the many other available forms.

In fact, there is a product which will strike many as an odd novelty but which is extremely popular and valuable to the health-minded. It is well known as nourishing food in Europe, and goes by the name "rose hips." Why they are called "hips" is obscure, but whatever the name, they are the fruit of the rose, which mature after the petals have fallen. After all, our selection of which of nature's products to eat or not to eat is rather arbitrary. For many centuries, tomatoes and other modern stand-bys were deemed poisonous. So now we learn that whereas oranges contain only 49 milligrams of vitamin C per 100 grams, certain varieties of rose hips contain from 2,000 to 6,900 milligrams of vitamin C per 100 grams.

Mr. Rodale tells us that the smoking of one cigarette uses up 25 milligrams of vitamin C from the blood. This is the amount of vitamin C in one medium-sized orange. On this basis, someone who smokes twenty cigarettes a day would have to eat twenty oranges that same day just to keep his supply of vitamin C at an even keel without allowing for any of its legitimate uses. In other words, you may be one of those who does not mind the possibility of cancer from smoking, but no one can take in stride a continual deficit of vitamin C.

Is bread your staff of life? Realize that the wheat crop comes largely from the Dust Bowl. What quality of soil can you expect? The wheat grows, but has nothing like a sufficiency of needed vitality. Then we use chemical fertilizers which deposit caustic, corrosive chemicals in the soil, destroying what was left of its natural values. As if that were not enough, we spray poisons and insecticides to kill weeds and bugs. Then the farmer harvests the wheat before allowing it to ripen in the field, and this gives him additional trouble. He has to give it various treatments to prevent spoilage and make it easier to handle. One of these treatments is cyanogen gas, which is such a dangerous poison that he has to wear a gas mask while applying it. The wheat, if such it can still be called, is then ground up fine in the steel-roller process. In the old days grain was ground in stone mills which revolved slowly and kept the flour at a low temperature. The present system heats the flour fast and removes a large amount of the valuable nutrients. The flour that came out of the old-fashioned mill seventy-five or a hundred years ago contained about 75 per cent of the vitamin B-1 of the wheat. Your white flour today probably retains only 10 per cent of it. The difference is what is destroyed in our modern milling and refining process.

Carrying on this sad saga, in the milling process the outer coat or bran is removed. This is a very important part of the wheat seed, containing large amounts of vitamins and minerals, especially iron and phosphorus. The outer bran contains good proteins. It is therefore set aside and sold as a separate product to farmers who feed it to their pigs, while the flour mill makes more money, and what is left for man is the inner part of the seed kernel, which is mostly starch and gluten and is good for making paste. It should properly also contain the wheat germ, but as we shall see, something is done to that too. As *The Health Finder* wisely comments, it would certainly be fairer to the public if it were placed at least on the same level with the pig. Actually the separation of the two elements of the grain creates nothing. The pig gets the bran and man gets the white stuff. Why not give both man and the pig their proper share of the whole grain? Why should pigs become healthy at the expense of the ill health of man?

In the next stage of modern bread-making, the wheat germ itself is removed from the wheat! In the human being this would be equivalent to removing the heart, for the wheat germ is the very heart and life of the seed, rich in vitamins and minerals. It is one of the richest sources of vitamins B and E and contains valuable proteins and fats. The vitamin E is usually sold to farmers to make their horses more fertile, and some people buy it for themselves. The drug industry makes millions out of this wheat germ. The public therefore must take vitamin E to replace the lost wheat germ from the bread it is eating. It is suggested that all of us be sure to take wheat germ, brewers' yeast, and bone meal to replace the valuable vitamins and minerals lost in the removal of the bran and wheat germ. Most of the food supplements are obtainable in easy-to-take capsule form.

Next, the flour is bleached, and all bleaches are poisonous to the human system. After all this, the bread is "enriched" with synthetically manufactured vitamins and some iron. In other words, some twenty or more natural vitamins and mineral elements are removed, and only four or five, mostly coal tar, are put back. Even those who pat themselves on the back because they consume whole-wheat bread should not be complacent, since most whole-wheat bread is treated with poisonous preservatives because it tends towards more rapid spoilage than the white breads.

Well, the story of bread could go on still further, but

you get the idea. Is it any wonder that Mr. Rodale himself eats no bread or flour products whatever, for this one-time staff of life is today a feeble and treacherous reed indeed?

All that is needed is enough purposefulness and will power to take advantage of the wonderful information and resources available in this country. There is little sense in the soft flabbiness and half-sickness of most of us in this land of plenty (while the rest of the world starves).

Fifty years ago it was different. At that time pure food was an unknown concept and ignorance about nutrition was the order. In his book, ***Business the Civilizer,*** E. E. Calkins cites the story of a deacon of the last century who was a grocer and lived over his shop. At the end of the day he leaned over the stairs to his helper in the store, and this conversation followed.

"John, have you sanded the sugar?"

"Yes, sir."

"Have you watered the cider?"

"Yes, sir."

"Have you dampened the tobacco?"

"Yes, sir."

"Well, then, come upstairs now for prayers."

VII

What's in a Name?

In early times the first or Christian name of a person was the only one that was recognized by law. While there were surnames, these were only words of description to identify one person from another having the same Christian name. The expression "Christian name" was the result of "christening," but before that the same thing was called a given name. There has always been some need for direct address, and a name is as old as Adam himself. Many years before surnames were adopted, the use of sobriquets had become customary. These names were generally founded upon some personal trait or incident in the life of the person, or referred to the place of birth. Alfred the Great, Friedrich Barbarossa (Red Beard), William the Conqueror, are some of these. But they applied only to the individuals and died with them, so they did not have the use of family names.

There were few surnames in England before the Norman invasion of 1066, but at that time the nobility began to call themselves after their ancestral seats, and it became customary to think that a person of rank should bear a surname. The noble or landowner used the word *of*, *von*, or *de* followed by the name of his estate, and thus established a family name. After the Reformation in England the introduction of parish registers helped to give permanence to surnames, and there came a real need for them among people of all classes.

For the sake of greater written accuracy, it followed that in any document not only should a man's name be inserted, but his business, his city, or some other definite characteristic. Thus John the Carpenter became John Carpenter. Heinrich of Berlin became Heinrich Berliner. Louis the Short became Louis Little. If there were no distinguishing characteristic, one was simply the son of his father, as William, his son, or Williamson, which might be abbreviated into Wilson.

The names still with us today denoting occupation are infinite, such as Smith, Baker, Fowler, Clark (clerk), Taylor, and many others. Sometimes a family engaged in weaving would take the name Webb.

Similarly the names denoting simple descent are evident in every nation and in many varieties. Sometimes this was done in the possessive way: to designate William, son of Robert, would be William Robert's, or, as now written, William Roberts. In Scotland the prefix *Mac* had the same meaning; and in Ireland, Tom, son of Hara, became Tom of Hara or O'Hara. The Norman *Fitz* is a corruption of *fils*, a son, as in Fitzgerald, son of Gerald. In English names there are countless Johnsons, Jacksons, Thomsons, and others; and until recently the Welsh adhered to the use of the word *ap* to show the same thing. David the son of Howell was known as David ap Howell. To this curious custom we owe the names of Pritchard, Powell, and Price, which are modifications of ap Richard, ap Howell, and ap Rice. The Germans have their Mendelssohns and the like. As far back as the ninth century the Saxon suffix *ing* denoted descent, as in the name Egbert Edgaring, and to this origin we attribute such names as Browning.

Sons took their fathers' names as early as the days of Priam, and their suffix was *ides*. Thus we have Greeks called Aristides, Thucydides, Euripides, and others.

A very complete system of nomenclature was had in Rome. Each citizen had three names. The first or *praenomen* was the given name. The *nomen* or name proper marked the name of the *gens*, or clan, for which we have no equivalent, but which was a large group of which the family was a smaller division. The *cognomen* was the family name.

Thus Publius Cornelius Scipio belonged to the Cornelian *gens* and the family of Scipio. Sometimes a fourth or *agnomen* was given, generally in honor of some military success, so this person was called Publius Cornelius Scipio Africanus, to denote the particular scene of his success.

This system was ignored during the Dark Ages after the overrun of Rome by northern peoples, so western Europe began all over again to create family names, as set forth in the first part of our account.

But in this chapter I shall use the word "name" in a broader sense than that of identifying people. The names we use for things are of even greater interest. Throughout this book I give frequent stories of the origin and development of certain words into their present meaning, but here I shall elaborate a little on the countless ways in which ingenuity and mother wit are revealed in language. And it will be noticed that no less wonderful than the stories in the words themselves is the ingenuity of the philologists who have traced these stories.

Without language, man could not be a social being. But of course he was not furnished with his first dictionary of words ready-made. Names he had none, but he did have the power of naming. He was not taught words, like a parrot, from without, for he was not a speaking machine. God gave him a capacity, the use of which was evoked by experience and necessity.

Language is the exact counterpart of the intellect. For every concept there is a word or words, and for every word there is a concept. Neither can exist apart from the other. Language not only fits the intellect by covering it completely and following all its turns, but also fits it in the further sense of not going beyond it. No savage has words for the typical things of civilization. The word comes only with the idea.

Explorers and missionaries have given us interesting examples of this thought. There is, for example, a Brazilian tribe, as reported by a Jesuit missionary, that possessed no word in the least corresponding to our "thanks." But what wonder, if the feeling of gratitude was entirely absent from their hearts, that they should not have the corresponding word in their vocabularies. Though inveterate askers, they never showed the slightest sense of obligation or of gratitude when they obtained what they sought, saying only, "This will be useful to me," or "This is what I wanted."

Again, in a part of Tasmania there are four words to express the taking of human life, one to express a father's killing of a son, another a son's killing of a father, and other varieties. Is it a wonder that a race so plentifully provided with expressions for murder is entirely wanting in the idea of love, as shown by the absence of any word to clothe it?

Individual languages reveal a great deal, but even more awaits us as we are able to trace the relations of words in one language to another, and these in turn to a common ancestor. To this has been given the name comparative philology. It is probably due in a scientific way to Sir William Jones, who as early as 1786 was attracted by the possibilities of Sanskrit, the sacred language of ancient India, as a probable parent language from which other tongues were derived. The German scholar Franz Bopp, working on this idea, became the founder of scientific philology of the Indo-European languages, and thus of comparative philology in general. A whole group of German scholars added to and supplemented those early researches. Bopp lived to see the results of his labors everywhere accepted and his name justly celebrated. Though he died a poor man, his genuine kindliness and unselfishness, his devotion to family and friends, and his rare modesty endeared him to everyone. A memorial fund was established in his honor fifty years after the publication of his first work, to stimulate researches along the lines he had inaugurated.

By tracing words according to this system, philologists have been able to reveal for us almost the whole state of various cultures. A striking group of words illustrating how this is done is that dealing with parts of vehicles. Not only the word "wagon" but the allied words "wheel," "axle," "thill," "lynch," and "yoke" appear in cognate forms in various Indo-European languages, indicating the development of the vehicle in the primitive stage of Indo-European life. But among these cognate words, conspicuous is the absence of a word for "spoke." This affords ground for the inference that the primitive wheel was solid in form and without spokes.

Whatever object the mind is especially turned toward, it throws out words often with marvelous profusion. When, some thousands of years ago, Sanskrit was still a living language and the sun and fire were looked upon as especially sacred, if not actual gods, fire, instead of having a very few names as now, had thirty-five, and the sun had thirty-seven, Max Müller tells us. But much more remarkable examples are those drawn from Arabic, as for instance the eighty names for honey, the two hundred for serpent, the five hundred for lion, the thousand for sword, and 5,744 words all relating to the camel, these being subjects upon which the Arab mind is strongly and persistently bent.

What would you think if someone wrote you a letter and signed it, "Yours without wax"? Foolish as this sounds, it is what you yourself say whenever you use the expression "Yours sincerely." The word "sincere" is derived from two Latin words, *sine*, without, and *cera*, wax. But there was a reason for the origin of this meaning.

In early times, man thought first on concrete things. Going always from the known to the unknown, such abstractions as he could master were founded on his experience with tangible, visible objects. Many words that we use today to denote abstract ideas are words of this kind whose meanings have been enlarged or expanded by figurative interpretations. This gives great picturesqueness to our language.

The word "sincere" is one of these. When the old-time potters found themselves with a piece of pottery that they had made imperfectly, they tried to conceal the defect with wax. If their work was perfect, it would be without wax. Hence the development of the concept that whatever was genuine or free from deception was without wax, or sincere.

Again, the word "scrupulous" is from "scruple," a small, sharp stone, as in a man's shoe. A person thus afflicted naturally proceeds carefully. Scruples are "small" considerations and indicate exactness. Our word "supercilious" comes from *super*, over, and *cilium*, eyelid. It means raising the eyebrows in disdain and thus expresses contempt or haughtiness.

"To ponder" means literally to weigh; "to ruminate" refers to the cud-chewing of cows; "brood," to meditate, goes back to the hen sitting on her eggs. "Perceive" and "comprehend" mean literally to seize and to grasp. "Obvious" comes from *ob*, against, and *via*, way; if a thing is against one's way, it is evident. "Circumstances" are things standing around.

Many of our common words are related to astrology. We have "ill-starred," "saturnine," "jovial," "mercurial," "lunatic," and others less obvious. "Disaster" is from *astron*, Greek for "star," and *dis*, negative, meaning a blast or stroke from an unfavorable star. Latin for "star" is *sidus*, from which we get "sidereal" and our word "consider," which means to observe the stars—supposedly from the practice of taking auguries.

A large group of words have their origin in the agricultural and other industries of the people. "Pagan" comes from *pagus*, a village, and signified merely a peasant or one who dwelt in a village, as distinguished from the inhabitant of a town or city. When Christianity was introduced, it became established first in the cities and centers of learning, so that in many instances, after the principal towns of a locality had become Christian, the neighboring villages remained in ignorance of the new faith. Thus the term "pagan" or "peasant" was used as a synonym for one who had not embraced Christianity. "Heathen" came into use the same way, signifying one who dwelt on the heath and also remained in ignorance of the new faith.

In the days of the Romans, when a general was to be honored on account of some victory in battle, he was given a feast for which a sheep was specially sacrificed. Latin for "sheep" is *ovis*, and from this is derived the word "ovation."

"Bachelor" is from *baccalarius*, a farm servant, a cowherd. It originally meant the holder of a small farm, presumably unable to support a family, hence an unmarried man.

Another farm word is "caprice." This comes from *caper*, a goat. The jumping about of a goat, goatlike capers, are capricious. The word "neighbor" is from *boor*, a farmer, and *nigh*, near, a farmer who lives near. "Companion" is from *panis*, bread, and *com*, with—namely, a bread-sharer. "Panic" is from the pagan god Pan, who capered about wildly; therefore we get the meaning of frenzy.

"Ambition" in its early days meant merely to go around. It came to mean the seeking for preferment because the candidate walked around seeking friends or votes.

Rivals were those who dwelt on the banks of rivers. After an inundation which destroyed their boundary lines, they contested their claims.

Roman cities had walls for protection, and were often built on high places. "Suburb" comes from *sub*, under, and *urbs*, the city, and refers to the building of houses under a city wall beyond the confines of the city.

Our word "idiot" was originally a term used by the Greeks to distinguish a private citizen from one who held office. It had no bearing at all upon his mentality. The word acquired its present significance from the assumption that only men of intelligence and ability were capable of holding office, and accordingly "idiot" came to mean one who lacked those virtues.

The word "person" was at first from *persona*, an actor in a play, named from his mask with large mouth for the sound to pass through, from *personare*, to sound through. After being a character in a play, it came to be any character, or rather that superficial part of him which gets in front of the mask but does not necessarily represent the true character. The personality may be a kind of make-up or disguise—an effect to give deliberate impressions not necessarily truthful, as opposed to the real character of the individual.

In lighter vein, let us look at the word "farce." *Farcire* is the origin of the term and was an expression of Roman cooks, meaning to stuff as with flour, herbs, and other ingredients. Whatever was stuffed in this way was therefore made up largely of things that, while palatable, were unsubstantial and designed simply to add zest and flavor. When we realize that a farce is a comedy with little plot or substantiality, but stuffed with various kinds of humor, wit, spice, the name seems singularly apropos.

At the other end of the scale is an even more picturesque word, that synonym for affliction, sorrow, anguish that we call tribulation. The old Roman farmers used a *tribulum*, which was a thrashing instrument or roller, for separating the corn from the husks. *Tribulatio* in its first meaning was the act of this separation. But some early writer appropriated the word and image for setting forth a higher truth. Since sorrow, distress, and adversity were the means for separating in men whatever was light, trivial, and poor from the solid and the true—their chaff from their wheat, so to speak—this writer bestowed the name "tribulations" to such trials. They were the thrashings of the inner, spiritual man without which there could be no fitting him for the heavenly garner. Thus a single word is often a concentrated poem.

As conditions change, these survivals in words often come to queer usages. In today's speech there are many words carried over in this way. Thus we no longer use sailing vessels, but our steamers "sail" at certain times. An airplane "lands" on the sea. We have weekly "journals," literally weekly dailies. A "manuscript"—strictly, written by hand—may be typed. Indeed, the ramifications of a study of our words take us into every age of history and every land.

Everybody knows that the original Celtic language of England was modified by conquests by the Romans, the Danes, the Teutons, and the Normans. In addition, as England grew prosperous and became a sea-going nation, influences were imported from all the lands she visited. Thus we English-speaking people have borrowed from all tongues, we have modified foreign words in the process of evolution, and individual men have deliberately created new words. Of all these influences, that of Rome was greatest. The conquest by Latin of European languages is almost as far-reaching in its effects as the Roman conquest of the ancient world. Latin itself had come into contact with many different languages spoken in early Italy and was already a rich and varied tongue, made up of Greek, Egyptian, and even more remote factors. Through her colonists, merchants, soldiers, and public officials, Rome carried her language everywhere. It was taken up by the unlettered natives who in dress, customs, and speech thought it desirable to make themselves as Roman as possible. In the countries now known as France, Switzerland, Spain, North Africa, England, and elsewhere, the old national tongues were practically abandoned for the new tongue. The decline of the Roman Empire did not interfere with the progress of Latin in Western Europe. It became the basis of the so-called Romance (from Roman) languages, now French, Italian, Spanish, Portuguese. These owe their variation from one another largely to the fact that writing was so infrequent that there was no standardization; each developed along its own lines, and was subject to different subsequent influences. English, as we saw, had its other invasions after the Roman, so that a combination of events accounts for our modern language.

The following words are the merest sample of what we owe to other tongues. In some cases they are directly borrowed, in others they have been modified.

LATIN: suggest, introduce, creed, confidence, virtue, spirit
GREEK: apotheosis, acme, biography, myth, physiology
ITALIAN: balcony, archipelago, cartoon, concert, piazza
SPANISH: alligator, contraband, embargo, ranch
PORTUGUESE: cocoa, caste, banana
FRENCH: justice, war, burlesque, apartment, automobile
PERSIAN: chess, peach, divan, bazaar, borax
SLAVONIC: polka, vampire
HEBREW: amen, Sabbath, balsam, cherub
ARABIC: admiral, algebra, alkali
MALAY: amuck, gong, junk
AUSTRALIAN: kangaroo, boomerang
POLYNESIAN: taboo
AFRICAN: oasis
NORTH AMERICAN: toboggan, wigwam, tomahawk, moose, hominy
MEXICAN: chocolate, tomato
WEST INDIAN: canoe, potato, hurricane, maize
SOUTH AMERICAN: alpaca, quinine, tapioca
DANISH: call, aloft, they, knife, sky, scare, egg, scald, bull

Other words have been appropriated from the national characteristics of various races. Thus the Germans are philosophic, the French fashionable, the Italians artistic—not exclusively, but dominantly. For example, the important place that Italy occupies in the development of music is to be found in the many Italian expressions prevalent in the common musical language of the world. Not only are the words "opera," "libretto," "oratorio," and other general designations adopted from the Italian, but practically also all of the musical terms are of Italian origin. *Soprano, alto, allegro, adagio, largo, crescendo, andante, finale, forte, staccato*, and many others might be mentioned. Such words have become almost as much a universal language as music itself, and make a common linguistic ground that has helped me out of many a difficult conversational place in my wanderings abroad.

Many special vocabularies of other arts and sciences may also serve this purpose. Who shall say but that these shall form the nucleus of the much-considered universal language of the future? But the idea of a universal language must develop in its own time, if it will. Whether a separate artificial tongue must be devised seems doubtful. There is too much hidden wisdom in the story of our words to forfeit the benefit by even such an efficiency device as simplified spelling, which is so largely the basis for unifying tongues. With our ease of intercommunication and our willingness to understand others, the instrument of speech must, by men's determination and ingenuity, solve the means in due course.

That words in their present form contain endless stories of moral and historical truth, as well as of passion and imagination, has been very thoroughly shown in the work of Richard Trench, Archbishop of London in 1851. As we would expect from an archbishop, his explorations into word histories have been rather largely directed toward the moral lessons that could be drawn from them. But so engagingly has this been done that one cheerfully offers one's self a subject for his little sermonettes, which are rewarding in their insight into life and manners.

Of words originally harmless, many have assumed harmful or unworthy secondary meanings, he says. "Officious" had reference to offices of kindness, not of busy meddling; "moody" was that which pertained to a man's mood without any gloom or sullenness implied. In "crafty" and "cunning" there was nothing of crooked wisdom, but only knowledge and skill; "craft" indeed still retains the more honorable use, a man's craft being his skill, and then the trade in which he is skilled. And think you, Trench continues, that the Magdalen could have ever given us "maudlin," in its present contemptuous application, if the tears of penitential weeping had been held in due honor in the world? Again, "tinsel," from the French *étincelle*, once meant anything that sparkles or glistens. Thus cloth of tinsel would be cloth inwrought with silver and gold; but the sad experience that "all that glisters is not gold," that much which shows fair and specious to the eye is yet worthless in reality, has caused the word imperceptibly to assume the meaning which it now has; and when we speak of tinsel, either literally or figuratively, we now always mean that which has no reality of sterling worth underlying the glittering show which it makes.

This leads Trench to inquire how it comes to pass that we do not speak of the retaliation of benefits as often as of injuries. The word does but signify again rendering as much as we have received; but this is so much seldomer thought of with reference to benefits than wrongs that the meaning has contracted. And what a mournful witness for the hard and unrighteous judgments we habitually form of one another lies in the word "prejudice"! The word of itself means plainly no more than a judgment formed beforehand, without affirming anything as to whether that judgment be favorable or unfavorable to the person about whom it is made. Yet so predominantly do we form harsh, unfavorable judgments of others before we have knowledge and experience of them that "prejudice" is almost always taken to signify an unfavorable anticipation about one, and "prejudicial" has actually acquired a secondary meaning of anything which is unfair or injurious.

His analysis of the word "passion" is also interesting. We sometimes think of the passionate man as a man of strong will and of real though ungoverned energy. But this word declares to us most plainly the contrary; for, as a very solemn use of it declares, it means, properly, "suffering," and a passionate man is not a man doing something, but one suffering something to be done to him. When a man or child is in a passion, this is not coming out in him of a strong will, or a real energy, but rather the proof that for the time at least he has no will, no energy; he is suffering, passive, not doing; merely negatively letting his anger, or what other evil temper it may be, lord it over him without control.

There is a lesson in our present use of the word "libertine." At one time it signified a speculative freethinker in matters of religion and in the theory of morals or government. But as, by a sure process, freethinking does and will end in free acting, so "libertine" came in two or three generations to indicate a profligate, especially in relation to women.

Once more, let us consider the word "kind." We speak of a kind person, and likewise of mankind, and we seem to be using quite different words. It is interesting to know that they are connected by the closest bonds. A kind person is a *kinned* person, one of kin—one who acknowledges and acts upon his kinship with other men, confesses that he owes to them, as of one blood with himself, the debt of love. And so mankind is *mankinned*, and we declare a sense of brotherhood with everyone when we use the word.

From this brief glimpse into a few word histories, let us consider some of the uses and powers of words. Clarity of thinking is always increased by the use of the correct word. Since our thinking is frequently so vague a matter, consisting merely of impressions and generalities, hardly thinking at all, it is often of help to write down one's thoughts, thus compelling the use of specific words and making the idea concrete if it seems of value. While communication between men was the incentive for language, the improvement of thinking as a result is fully as important. Since words are the tools of thinking, the larger the vocabulary, the clearer and more exact the thinking. Most executives in any field are notable for their good vocabularies.

Confucius once was asked what he would do first if he were to administer a country. "It would certainly be to correct language," he replied. His listeners were surprised. "Why?" they asked. The Master's answer: "If language is not correct, then what is said is not what is meant; if what is said is not meant, then what ought to be done remains undone; if this remains undone, morals and arts will deteriorate; if morals and arts deteriorate, justice will go astray; if justice goes astray, the people will stand about in helpless confusion.

Hence there must be no arbitrariness in what is said. This matters above everything."

We obviously cannot give a word or name to a thing that we do not recognize. But to recognize it is to give it a place in our thought. To think of it is ultimately to express the thought in some kind of action. Hence, what we recognize in a way specific enough to name has a very direct bearing upon our behavior.

Certain primitive people believe that to know a person's name is to have power over his soul. They are careful to conceal their own names from strangers. The exchange of names as in our introductions is a mark of confidence.

Since naming is first mentioned with reference to Adam in Eden, it is of interest to consider this more fully, for the biblical stories often have a symbolic meaning that gives plausibility to the fanciful language in which they are couched. The symbolists tell us that the biblical injunction to Adam to name all the animals means that Adam, representing man or humanity, must recognize the traits represented by lion, fox, lamb, snake, etc., in his own nature and surroundings. Man is given dominion over all these kingdoms only as he names, that is, recognizes, controls, and gives them their true place.

Elaborating this opinion, the body is the Garden of Eden given each of us to cultivate and rule. It is, according to *The Truth About Evolution and the Bible*, by Curtiss, filled with all kinds of plants and trees (nerve trunks and their forces, also fruits of the spirit) which the self must tend, cultivate, and ultimately bring to perfection. It is also swarming with "every fowl of the air" (thoughts and ideas) and "every beast of the field" (appetites, passions, and desires) which the Lord (Law) brings to the attention of man to be recognized as being within and a part of himself, and to be named, controlled, and trained before he can re-enter and ultimately become the ruler of Eden (have perfect control of himself and his powers).

This interpretation is at least more interesting than

the plain surface statements. Whether we accept the symbolic story or not, the naming and recognizing of things is a distinctly creative use of speech, a power whose significance we do not begin to realize.

The word as a creative force is known to everyone who has even a bowing acquaintance with biblical literature. The book of Genesis tells how all things were called into existence by a divine utterance. God said: "Let there be . . . and there was." John says, "In the beginning was the Word, and the Word was with God, and the Word was God. . . . The Word was made flesh."

So, then, if we will accept this, God spake the word and the world was created. Man, made in His image, speaks one little word, and a new world of gladness is created in someone's life, or he speaks another word and someone else is disheartened or ruined.

The creative word is given countless instances of authority in the sacred writings:

> Man doth not live by bread only, but by every word that proceedeth out of the mouth.—Deuteronomy 8:3
>
> He that keepeth his mouth keepeth his life: but he that openeth wide his lips shall have destruction.—Proverbs 13.3
>
> A soft answer turneth away wrath.—Proverbs 15:1
>
> Death and life are in the power of the tongue.—Proverbs 18:21
>
> Speak the word only, and my servant shall be healed.—Matthew 8:8
>
> But I say unto you, That every idle word that men shall speak, they shall give account thereof in the day of judgment: For by thy words thou shalt be justified, and by thy words thou shalt be condemned.—Matthew 12:36,37
>
> The sower soweth the word.—Mark 4:14

All this sounds like religion. But there is much in the Bible besides religion. There is the history of people like ourselves, the story of their problems, their effort to understand life, as we also try to understand it, their ways of adjustment, their search for betterment, for happiness, the cause for ingenuity then as now. We may not care for the purely religious aspects of the Bible, but we may have a tolerant ear for words setting forth the experiences of people who believed they were in close contact with fundamental principles.

Science is coming to our rescue, too, in unfolding to us the significance of these warnings as to the use of words given with so much conviction and authority. Science is telling us now that the spoken word can never die, but will go on working out the potency and power inherent in it—hence the importance of the admonitions and my justification for dwelling on them.

With such developments as radio and television, we are more open-minded about scientific possibilities that previously we might have dismissed as too fantastic for acceptance. We are therefore not too greatly surprised to learn that it is being shown that every word and tone spoken earnestly by man is registered in the ether records, those infinitely fine, subtle waves that are part of our surroundings and which we are using in so many scientific exploits today.

We are familiar with the teaching that matter is indestructible. Our words, carried into the ether, are matter of a very subtle but powerful kind. Just as electricity is not visible, so with some of these other mighty powers that we are just beginning to learn about. Until our words are therefore either fulfilled or neutralized by their opposite, it would seem that there they remain, moving toward their fulfilment. With them we create either what we want or what we do not want. Thus to utter the harmful, pessimistic, or unkind thought is not merely an unsocial thing but one of great personal jeopardy scientifically. That thoughts themselves are actual entities is scientifically proved by such investigators as Elmer Gates of the Smithsonian Institution, by Boirac of the University of Dijon, whose works on *Our Hidden Forces* were rewarded by a great prize from the University of Paris, and others, including the work of Kilner of London in actually photographing thoughts. The evidence for this appears in a fascinating book by Leadbeater and Besant, called *Thought Forms*, and containing over fifty photographs in color, of different types of thought. The definiteness of these makes the reality of thoughts a convincing fact and takes it out of the realm of mere vague fancy.

To utter the word, then, is to put thought into action. The power of the spoken word is set forth in the findings of Henri Bergson in his *Mind and Energy*, Ouspenski in *Tertiam Organum*, Sir William Crookes, and others. Here is not the place, however, to present proofs at length, but it may attract us enough to practice it without the proofs, for there are such good precedents for the belief. In any case, it would seem wise to err on the side of caution and to waste no such powerful force in frivolous or negative chatter, but to speak rather only positive words, helpful, strong, kindly, and constructive. If one is wanting ingenuity to create conditions that he desires, here is one thing that everybody can do. Until we can speak words of health, strength, cheer, courage, and love, which are the gist of the sacred teachings referred to, we had better follow the injunction, "Let your communication be, Yea, yea; Nay, nay: for whatsoever is more than these cometh of evil."

What are the most beautiful words in the English language? What are the words most used by our poets in writing their songs of love and beauty? There are certain words of peculiarly dynamic quality that are able to change one's mood from dullness or melancholy into a more acceptable frame of mind. Such words not only strike a musical chord in our mental reservoir, but also seem to stimulate us and exert a warming, elevating influence over our lives, inspiring us to greater effort and achievement. Such words do not require a context. They are full of meaning and inspiration just by themselves. Each of us can discover words of this kind that make a special appeal to us personally and serve this valuable purpose. My own list would include words like *radiant, vibrant, sun, star, buoyant, hearty, peace, love, gentle, clear, nobility, trustfulness, confidence, vigor, creativeness, power, glow, hope.* "Whatsoever things are honest, whatsoever things are just, whatsoever things are pure, whatsoever things are lovely, whatsoever things are of good report . . . think on these things," says Paul (Philippians 4:8). These words are impersonal, and they help us to lay aside the perplexities of personal affairs. When you think of any of them, it cleans the mind of undesirable and confusing things. Each one seems to me both restful and stimulating.

Does this seem a long way from how we got our names? But what is the use of having names for things if not to use them so as to effect the largest results? It is for this that we have our mother wit, and our use of the creative word is something so simple and so practical that I have dwelt on it at length as something we can, if serious, practice with excellent result.

One of the most beautiful sentences in our language is "I love you." A man quoted in *Popular Knowledge* once took the trouble to find out how this is said in other languages besides English. I do not attempt to verify it, but even in its present state the list may prove useful in some emergency. Here it is:

ITALIAN: Vi amo.
FRENCH: Je t'aime.
POLISH: Kocham cie.
GERMAN: Ich liebe dich.
ARMENIAN: Yes Kee Seerem Kez.
JAPANESE: Watakusiwa Anata Suki Masu.
PERSIAN: Chouma ra Doust Darem.
DANISH: Jeg holder of dem.
HEBREW: Ani ohev osoch.
EGYPTIAN: N'achqeb.
PORTUGUESE: Eu Vos Amo.
RUSSIAN: Ia Vas Lioubliou.
SPANISH: Te Quiero.
SWEDISH: Jag Tycker om Eder.

HAWAIIAN: Nui Kouou Aloah No Oe.
DUTCH: Ik Bemin U.
HINDU: Main Tym Ropijar Karyn.
GREENLANDISH: Univifgssaerntdluinolerfironajungnarsigujok.

This way of saying "I love you" in Greenland must be the reason why the Arctic nights are so long!

VIII

From Cave to Skyscraper

CHILDREN were the cause of architecture. Nature-man could get along fairly well with rude makeshifts; but the helpless child in its infancy was no fit companion for his hardy life. The hunt, the fight, made no place for children, and their mothers too were rather unable to be strenuous at all times. Gradually it became evident that the nomadic style of living did not adapt itself to all, so the women and children were left in caves and other places of partial safety, the men returning after the hunt. And thus a home was devised to protect the child against the heat and the cold and the animals that came to prey.

Man always had the example of nature to follow. He saw that the animals satisfied their hunger with certain kinds of food; he did likewise. He saw that they lived in dens or caves or burrows or nests, according to their kind. Again he took the hint. Just as today we work from the known to the unknown, even so did the ingenious primitive. His basis for experiment was always the material at hand. With that he could make new combinations. Had there been no animals, none of the furnishings of the earth as they then existed, had man simply arrived on barren soil, it is doubtful whether any civilization or culture could have been developed.

To obtain food was the first urge of his being. To obtain shelter came to be the next most insistent thing in the world —the second great law of nature. These two make up the great desire of self-preservation, the will to live. From this are derived the progressive steps not merely of living but of dwelling in comfort, peace, security, and the other attributes of civilization.

Whether we find a palm-shaded hut in the languorous tropics, with the inmates sitting on a bamboo mat, or an igloo in the Arctic, with the family sitting on an ice cake, it is there because someone wanted a home. But at one time home was not one of man's ideals. The primitive roamed the earth along with the animals, each seeking merely enough to eat and protection as it might be obtainable. Largely man lived alone because everyone was a potential enemy—a rival seeking for the same scarce treasure.

When the constant sense of fear became too depressing, man again applied his ingenuity. Some prehistoric modern announced for the first time the idea of coöperation! What a tremendous step forward it was when there came the realization that a dozen or more could live in one place together, more easily, safely, and comfortably than each alone! The result of this exceeded all expectation. Certain men, it was found, had special gifts which could be enjoyed by the group. One was clever with tools and weapons. Another was keen of eye, another good at skinning. So groups or tribes did form, the members agreeing to help one another against their natural enemies. The earliest model of a league of nations was established, and civilization became that day assured.

In those most ancient times, before "political influence" was brought to bear, the most able man became the leader. Instead of moving from place to place, men gained greater comfort by selecting a good spot and staying there. A good spot was one where they were fairly safe from attack, protected from storms, near to water, and close to good ground, the fertility of which they were beginning to use. Home, since those early ages, has been, according to the time and the place, the tents of Arabia, the huts of the jungle, the peasant homes and châteaux of France, the shacks and villas of Italy, the hovels and palaces of Russia, the humble dwellings and great castles of England, and greatest of all, to us, the American home which represents an average harmony, luxury, comfort, and convenience superior to any that has gone before. But long the road and steep the climb to that achievement.

The primitive father of tropic climates devised for his children a strange residence. Surrounded by luxuriant vegetation, the tree suggested itself as a home. The tree it was that fed and clothed and protected him. When we read of the endless uses and products of a coconut tree, for instance, we can see ample reason why it should seem sacred, a place of truce and peace. In its branches he found refuge from floods, insects, and beasts. He erected a scaffold within its limbs and fashioned its leaves into a thatched roof to ward off rain and sun. A ladder sometimes gave access. Here he could flee beyond the reach of beasts before he had acquired the means for making them do the fleeing. This seems pathetically strange to us now, with all our modern sophistication. Yet there are tropical countries where the tree house still survives. Some sections of the Philippines have them, and they are also found in Asia itself.

According to recent excavations, the earliest Greek houses were built of wood roofed with leaves and were circular in form, thus indicating their treelike ancestry. To enlarge the space the circle was later made oval, and the roof was supported by posts. After that it became rectangular, sometimes with one end curved, creating the prototype for the apse. Peoples roving from southern climates northward remembered their arboreal homes, and in some of the great halls of Norse chieftains a tree served as a nucleus, the hall built around it. In *Die Walküre* the hut of Hunding is built around a huge tree which extends through the roof. Into this tree Wotan had cast a magic sword which has special significance in the working out of the opera.

In the mountainous and rocky regions of the North, caves were preferred as popular shelters. A hillside opening might be enlarged and strengthened if not actually created by man. Neanderthal and Cro-Magnon men dwelt in these. It is said that a reversion to the cave dwelling may bc found in France, where some two million of people in the Loire

Valley have found emptied quarries that are free from dampness, warm in winter and cool in summer. The entrances are vined with flowers.

The prehistoric cave hut, however, used no flowers to make itself conspicuous. It still preferred low visibility. The door faced the west when possible, to avoid the biting east winds. The roof was pitched high to permit the smoke to clear above eye level. The sides were very low, and goats, sheep, and pigs could climb to the roof and graze over it, for it was covered with sod so that grass went on growing. The lines of the whole building were irregular, so that anyone would think it merely a mound of grazing ground.

From the mouth of the cave came light, wind, rain, air, sun, and all of good and evil. It must have been damp, dark, smoky, drafty, and generally unpleasant. For all that, it was probably not the discomfort of it that led to a change. After all, it offered considerable safety, and in those days one was not finicky or spoiled with too much luxury. It was doubtless the scarcity of caves that made ingenuity get busy again, for the people, as they exhausted the growth and supplies of their hillsides, would have to move on to open country where caves were lacking, and something else had to be done.

Building on the known, men made artificial caves by digging them underground and setting up ladders to reach into them. But some less conservative pioneer, in a time of peace, tried a cave above ground, and this was the right ancestor to our modern architecture.

Whatever the primitive builder knew of architectural form was derived from nature. Looking over the land, he saw the rounded hills and the curving horizon, the circular trees, the clouds. He discovered the curve. His first attempt at building was to construct a mound. But the mound was not so much a residence for himself or his family as a resting place for his dead chief. The French Revolution was millenniums in the future; liberty, equality, and fraternity were not yet the general sentiment. A ruler was a superior. Let the horde of slaves be the nonentities they were thought to be. Any shelter, or none, for them; what matter? But the dead chief must have a dwelling for eternity that was worthy of him.

Somehow the mound, wonderful as it was, seemed to lack in distinction. Its builder felt it was too much like the work that nature accomplished much more beautifully. He wanted to be himself, to do something nature could not do. The result was a new kind of mound, rising higher from the earth, with flat sides instead of curved, with straight lines and geometrically regular. In short, he had created the pyramid.

The pyramid, too, was a tomb, not a residence. It was believed that the Egyptian had a *Ka*, or double, which continued to live after the physical being passed away. In order to keep the *Ka* in necessary supplies in the other world, burial came to be a most elaborate affair. Everything that he used in his earth life was entombed, including furniture, jewelry, clothing, and all the other appurtenances of living. For this reason archeological excavations are specially informing when directed towards tombs, and such explorations as those of the tomb of Tutankhamen tell us more perhaps than any other available source about ancient Egyptian culture. From these marvelous tombs, built of huge stones, magnificent in their massiveness, we have derived the sense of solidity and power that subsequent building was able to develop.

This does not apply to the great and most famous Pyramid of Gizeh, which is the epitome of all that is ingenious. Built in the very center of the land surface of the earth, the great thirteen-acre pyramid was constructed by intelligences far beyond our present attainments as a prophecy for these times some six thousand years later. People have long been trying to fathom the mysteries of this pyramid, for that it contained remarkable secrets was evident when it was discovered that it was not a tomb and no one was buried there. Archeologists, astronomers, mathematicians, and other scientists have studied it intensely and many fascinating books have been written proving the various prophecies and revelations. The mathematics involved in its construction and its interior passages are symbolic in the highest degree and are utterly accurate portrayals of major events in the last six thousand years. The more that is discovered about this pyramid, the more it is realized that these ancient builders knew far more than we do today about everything under the sun. This is not the place to develop the fascinating story of this Great Pyramid. To do so would require a book in itself. But one cannot mention pyramids without at least this casual reference to this massive mystery in stone.

Though the pyramid grew into a form of complicated interior design, the cave and the mound developed more simply into a dwelling for the common man. The next step was the hall, which like its predecessors was one great room, lacking partitions into smaller rooms or stories. Our chapter on "The Struggle for Fire" tells something of its early development, and other references may be remembered (see Index). The entrance to the hall was more like a window than a door. The top was brought so far down that one had to stoop low to enter, and the threshold was so high that the one coming had really to climb in. This was a measure of protection to the inmates, for if you were an enemy, you could be readily overpowered while trying awkwardly to make an entrance.

A much more troublesome part of the hall was the window. As this was only a hole in the roof from which smoke could escape, it was very difficult to defend or protect. The master of the house was often killed from that point as he sat by his fireside. Long benches were fixed on either side of the fire. The middle of the long seat was raised above the level of the rest. This was called the high seat and was reserved for the master and mistress of the hall, as it was the most comfortable location.

At this period comfort was still so embryonic as to be

Pan American

The Sphinx and the Great Pyramid of Gizeh.

almost nonexistent. Cold winds howling through the many open cracks; fire that roasted the face and let the chills creep up the back; soot and smoke to sting the eyes and nose; no bathing facilities; hard benches; dim, flickering, smelly candles—the love of life had to be strong to endure it. Climate made it hard for these people, and their ways of adjusting to it were slow and weary.

But adjust they did, for climate is an insistent factor. In every land, at every time, things are done at the behest of climate and its results that are practically inescapable and inevitable. Climate, for one thing, dictates the materials of building. In Eskimoland the igloo, which is a mound house of snow, as the crossword-puzzler may recall, serves as a model of Arctic comfort. It serves as a home for two or three weeks, then a new one is built. Because it is new, it is clean and sanitary. It is warm and comfortable, because long experience has taught what proportions and thicknesses of snow to use that will stand up safely and keep the weather out. One candle gives as much illumination as three electric lights because of the intense whiteness of the snow. Here is a satisfactory adaptation to climate. But how long would even the finest igloo serve down in Ecuador?

In the forest lands, such as Switzerland and Germany, the chalet is of wood, and timbered houses are common. The German word to build, *bauen*, is closely connected with *baum*, a tree, so that the interrelation of climate is again evidenced. Stone and marble floors and walls which are agreeably cool in Egypt and Italy are unwelcomely chill in the North, unless plenty of thick rugs and bright hangings can be had to compensate. On the other hand, overhanging roofs, tiny, deep windows, and thick walls that protect an Egyptian house from the fierce glare of the African sun would give dampness and gloom in a place like London, which has a limited pale-yellow sun at best, and a surfeit of heavy fogs as a regular thing. Holland, too, bows to climate as influencing topography.

In Holland, where most of the land is reclaimed from

the sea, there is always the problem of infiltration of water despite every safeguard. The foundation of the Dutch house has to be pumped periodically, and the costliness of the foundation is so great that Dutch houses have been made tall and narrow for their size. The foundation must be as small as possible and must be utilized to the maximum. Steep, narrow stairways are the result of the narrow buildings, and since it is impossible to carry anything bulky upon them, each house has a crane projecting from the topmost gable by which heavy things are hoisted into the windows.

The results of climate are always especially noteworthy in the roofs of buildings. You have only to think of the high-ridged, many-dormered roofs of Alsace, designed to shed the heavy snows, and then, for contrast, picture the flat roofs of those houses in the tropics, where there is only the sun to contend with, and which the people use as roof gardens after sundown to get a breath of air.

But important as climate is in building, it is but one factor in the great work of architecture. The word "architecture" should only be applied to buildings that are raised by esthetic qualities to the rank of art as distinguished from purely utilitarian or mechanical structures. The climatic appropriateness will be taken for granted. We will not build a Swiss chalet on the plains, nor a granite house near the marble quarries of Carrara.

The mind naturally works up from the necessary to the beautiful. There had to be the idea of the hut before the idea of its beautification. The oldest existing monuments of architecture in this sense are the palaces of Chaldea and the pyramids of Egypt already referred to. The Egyptians also built very imposing temples of great size and majesty. Smaller temples were the chief architectural contribution of Greece, and the columns designed by Greek architects have entered into every type of subsequent building, and have carried the Greek influence far and wide in every land.

The Romans, too, were great builders and architects. We describe elsewhere their monumental baths. Their roads, bridges, and aqueducts will also be remembered. In these they gave their great gift to posterity, the use of the keystone arch, which made possible feats of strength in building that were infinitely simpler to achieve than in the ponderous Egyptian way. (It is understood, I hope, that the ramifications of the story of architecture are too labyrinthian for more than the merest sketch. This is not even an outline, but only perhaps a hint of a few of the outstanding developments.)

One of the most important structures of the Roman Forum was the basilica. This was the hall of justice, a long, rectangular building. Since the basilica was the largest hall that the Romans had, its form was copied by the early Christians for their churches. This gave it such importance during the centuries to come that it will be of value to describe it. The center was the large open space that we now call the nave. On either side of this were rows of pillars that formed aisles. The end of the nave was curved and formed a semicircular apse called the tribunal. From here cases were heard. A rail separated the tribunal from the body of the hall. This because of its openwork was called *cancelli.* It was a marble screen and constituted the bar to which the accused were brought. The pillars were in two tiers, the upper one enclosing a gallery reached by stairs either inside or outside the structure. The basilica was separated from the Forum only by open colonnades. It might be curtained off, but was generally open and used as a meeting place for idlers. It is amusing to see today, scratched in the flooring of one of the basilicas, a Roman gaming table.

When used as a model for the early churches, the bishop's throne was the outgrowth of the judge's seat. The altar of Apollo gave place to the communion table, the *cancelli* to the chancel, and the fountain in the court in front to the holy-water basin. And so the name was also handed down and given to Christian churches, though there is no church in Rome that was ever a pagan basilica or hall of justice. Many of the so-called basilicas are not true basilicas, for they have introduced the transept to give them the form of a cross. So again we see mother wit using old ideas in new forms.

After the decline of Rome and the growth of the Church, religious influence superseded every other in architecture. There came the great era of church construction. Cathedrals followed one another throughout Europe. In the Gothic types, which had their best expression in the North, marvelous structures were erected.

Medieval achievements in the realm of architecture were especially brilliant. We owe the Gothic style of architecture to the medieval builders who, without machines or steam engines, were able to erect tall cathedrals of marvelous beauty. Neither the ancient Greeks nor the Romans knew how to construct the slender spires, the graceful pointed arches, the high vaulted ceilings, the flying buttresses, the exquisite stained-glass windows, that one finds in the medieval churches and public buildings. Westminter Abbey, Notre Dame of Paris, the Rheims Cathedral, cathedrals of Cologne, Milan, Strasbourg, Rouen, and countless other splendid edifices still stand as matchless examples of the medieval builders' art and ingenuity.

The ground plan of a Gothic cathedral represents the cross, the circular window the rose of eternity. The immense heights, the vagueness of the distant recesses, the mysterious, subdued light, the heaven-reaching spires, all imply the symbolic uses of architecture.

In my chapter "From Echo to Static" I show the development of the medieval tower. This feature was a most important part of the castle of the Middle Ages—a structure that shows how buildings adapt themselves to necessities, and how man's ingenuity applies itself to adjustment to circumstances and conditions.

This was the feudal age, when knighthood formed the aristocracy and deigned to do nothing but fight. Everything was planned for defense. The castles were glorified fortifications, and every measure was taken to make their conquest difficult. The walls of the medieval castle were often ten or fifteen feet thick, with turrets and battlements and lofty towers pierced by small windows. Around the castle there was a deep ditch or moat. This was filled with water and was spanned by a bridge that could be removed by pulleys in time of attack. Whenever this drawbridge was raised, the portcullis, a heavy grating of wood, was lowered across the gateway, barring the entrance. Vulnerable doors and windows would be small and few, and there would be rounded corners which presented fewer points of attack than square ones. The top of the towers projected, having a gallery pierced with holes in its floor, through which could be dropped stones or melted lead or hot oil or arrows upon the enemy below. All roads leading to a castle would be made artificially hilly if they were not naturally so, for incoming enemies were at an additional disadvantage struggling up a hill. The actual living quarters faced the inner courtyard, as this was the most protected place.

The invention of gunpowder wrought a revolutionary change in architecture, in the whole idea of shelter, and in medieval civilization. For after this event, and the manufacture of cannon in the fourteenth century, the stone castle was no longer built, for no walls could be made so strong that they could not be battered down with cannon balls. The whole form of warfare changed. Knights in armor were helpless against gunpowder. Castles were no longer impervious to attack. It became necessary for central governments to supplant the rule of feudal lords. The King's Peace offered organized protection, thereby making it possible to build houses that were fitted for domestic life. No longer walls for defense, but walls merely for protection against the weather sufficed. Instead of slits for windows, broad panes of glass now admitted light. Instead of perching an important building on a high, protected spot, charming locations by a riverbank could be selected. Gardens became possible. Peaceful occupations ensued. Prosperity increased. Towns and cities grew rich and independent, without the overlordship of the nobles. Such wealth was accumulated that cities like Venice, perfect dreams of architectural beauty, came into existence.

Here the narrow lanes and waterways left little room for the people to take the air in the normal way, and that delightful invention, the balcony, was evolved to secure from above what was too difficult to obtain at ground level. The military necessities of the Middle Ages had prevented the development of anything so vulnerable. There were occasional prominent balconies erected on town halls and looking out on the public square, but they were different. Such a point of vantage was called the Ringhiera, and is somehow related to our word "harangue." From here the doge or the ruling governor would address the people, giving them the official news, or any public notices.

But for the sheer delight of a balcony it was left to Venice to excel. Here the climate was neither too hot nor too cold, and the town was prosperous and peaceful, so this charming feature was peculiarly suitable. In those parts of northern Italy that came under her influence, balconies abound in the finest examples. Verona, the city of Romeo and Juliet, possesses some excellent types, and it is easy to imagine many tender and impassioned scenes known only to the silver moon and to these balconies, where ladies hung to hear the sonnets of their cavaliers. Along the Grand Canal they are a great help to romance, giving the perfect touch to the perfect setting, affording a background of soft, rich color, a frame of delicate tracery, a consciousness of the rippling waters below and music in the air beyond.

Like Italy in the South, Belgium in the North at one time formed the commercial center of Europe, and because art comes on the heels of wealth, it soon developed great artistic importance. The wonderful "Grand Place" of Brussels is one of the finest public squares in Europe, its beauties showing the possible interrelation with daily business life. The fine buildings are not castles or mansions, but guilds, town halls, exchanges, belfries. The pictures are portraits of the solid and successful merchants, dressed in fine furs, jewelry, brocades, and the rich fabrics of older times.

The influence of occupation is seen in a very different though equally enchanting manner in some of the timbered houses of Germany. In Hildesheim, which is the Nuremberg of northern Germany, there are several hundred of these quaintly timbered houses, most of them dating from about 1500. These buildings take their origin from the farmhouse. The gateway must allow room for a laden harvest wagon to enter to the barns on the inner court. It is therefore about twelve feet high and proportionately wide. Only the ground floor was used for living purposes. An intermediate floor with but a low ceiling was used for servants' quarters, while above in the gables were granaries and storerooms where corn was kept. In many of these houses, the higher one goes the wider each story projects out, until at the top one can almost hold hands across the narrow streets. This economized space. The large farm wagons could pass by below, where room was left for them, but above there could be more house.

These houses are gay with color and are carved in symbolical ways to represent the trade of the owner, or the muses, apostles, seasons, gods, saints, or other subjects that lend themselves to decorative interpretation. Sometimes the treatment was facetious and humorous. This wealth of imagery in France was expended on great cathedrals, but in Germany it was lavished on buildings of the dwelling house type, with most engaging effect, for the well-to-do.

The day is gone when one needs great riches to possess the luxury of kings. Indeed, the traveler in Europe who visits at first hand the seats of royalty in many countries will be surprised to see how very much more comfortable and luxurious is an American home of today as compared with some of these grand relics of the past. The very size of some old palaces is against them—great, rambling, disjointed things that they are. Their thick walls and prison-like windows devised to keep out invaders, their hard stone seats, their altogether inadequate plumbing and utilities, their gloomy lighting facilities, their cold and drafty halls, were perhaps the best available, but in no sense would satisfy our simplest wants today. The modest home today presents by far a wholesomer and finer aspect than the drafty, cheerless, barely furnished palaces and castles of old Europe. What good were the gaily frescoed walls if the floors were cold and clammy and uncarpeted? What good were the great carved chests if one had to sleep on a boardlike bed without a spring or a mattress to bless itself with? Today the great tradition of the past gives us beautiful things to build upon, and our twentieth-century ingenuity has contributed to this beauty a comfort and sense of fitness heretofore unknown, made possible by our increasing wealth.

The coming of prosperity in all ages always meant better homes. It was not dearth of ideas but of materials or leisure that kept the low level. Egypt, Athens, Rome, Venice, Bruges—go back as far as you will: where there was peace and wealth, the ideas were present and could be carried out. And at other times the ideas were present, but makeshifts had to be employed. Man did the best he could under the circumstances.

How long the way from the rude shelter to the vast palaces, or even the merely decent homes of the everyday man—from the simple shrine to his great temples and cathedrals! The idea of protection has undergone many a change from the time when the primitive babe needed shelter, to the care now bestowed by man upon his gods, his dead, his animals, his tools, his business, his games, and his other comforts. Building goes ever forward. New materials, new modes of construction, are at work to lend their influence; the skyscraper would be impossible without steel and without the elevator. When we survey our skyscrapers and see the height which civilization is reaching, one factor is apt to be overlooked—the electric elevator. But for these cages rising toward the heavens, then returning at will to earth again, man would still be living in low-spreading structures, and the buildings of the population of a single city would be sprawled over the area of a state. The elevator made practical the erection of buildings upon small plots of ground whose roofs nose the clouds, and it has permitted cities to extend vertically as well as horizontally.

Plastics are entering into construction, decoration, and many new uses. "Plastics" is a general term, like the word "metals," or "fabrics," and embraces a large number of different substances. Their distinguishing quality is that they are man-made substances formulated in the chemist's laboratory. They are not found in nature. They are derived from various unexpected sources, such as milk, tar, waste products, and even air. They come in all colors from clear crystal to solid ebony. They can be had immune to acid or alkali, water-proof, fire-resistant, unstainable. Most are nearly light enough to float. They can be readily cut, bent, molded, sawed, carved. They can give the hard, glossy, smooth surface of glass without its breakability and coldness, or the warm look of old ivory. Most of all, they can be created out of inexpensive simple materials anywhere in the world and can be mass-produced in any shape.

Air-conditioning is giving us control of temperature within doors, and fluorescent lighting tries to rival the sun. Unit furniture, interchangeable, vari-purposed, often in metal-tubing effect, plus buttons everywhere to push for magical efficiency, mark the modern building. New social habits, new ways of living and thinking are all reflected in our architecture.

For in buildings are the records of the life of the community. Its social order, its religion, its economic condition, its tastes are all revealed. From small needs to great, from physical comforts to spiritual yearnings, our ideals move on, and buildings afford the place and means for their expression.

IX

Up From the Nude

NOT TO GO all the way back to the Garden of Eden, it is nevertheless true that many things that we take for granted today were first created in oriental minds. The very idea of dress, to which we are now so habituated, was a growth, and the evidence seems to point to the inference that it came about not so much as a result of climate or of modesty as for the urge to be decorative. Indeed, the word "garment" is probably contracted from "garnishment," and there are places where methods of paint or tattoo serve the purposes of clothing.

For modesty is largely a matter of habit. Costume is derived from custom. It is highly conventional, and certain things that are entirely proper at certain times and places are quite the contrary under other conditions. Ballroom attire in the business office, the bathing-beach costume in the city, and many other examples will suggest themselves. Among the nomads of the desert, the Arabian man always wears a cloth across his face which is never removed in company. Originally these cloths were put on to protect the face from the glare of the desert and the driving sand, but so ingrained did the habit grow that such a man regards it the height of immodesty to show his face even to his own family. Until recently, a Moslem girl in the street could lose her dress quite unblushingly, but to lose her veil would cause her to wish the earth to open and swallow her.

In Sumatra the natives go about entirely nude except for knee coverings. There appears to be no instinct in man to cover his sex organs. Scandinavians in the sixteenth century undressed in one another's company and, except for a nightcap, went to bed stark naked. They felt that men need not be ashamed of what God had created. In Japan, both sexes mingle in the bath, as was at one time the case in European public baths. Savages who are compelled to wear clothes feel the same embarrassment as civilized men might, were they obliged to discard theirs. The queen of an African tribe gave audience to Livingstone entirely nude. When a missionary sends such a person a skirt, he finds on his next visit that it is draped around the head as decoration. Thus modesty seems quite uninstinctive and merely a matter of habit or training.

Of course, an essential of clothing in the modern sense is the creating of the materials from which clothing could be made. Skins of animals, feathers, leaves, and grasses were all very well, but mother wit went farther than these. The remarkable nests which certain birds were building in the primeval forests gave to man the idea of combining fiber or hair into materials that could be used for clothing. Following it out, he first made cages for animals, nets for fish, and other things by interlacing vines or slender boughs. Then came baskets, cradles, and mats. Soon he was exploring for threads of finer texture that would not scratch the babe and that would yield themselves to greater versatility. By watching the spider, the art of weaving was inspired. When these materials were first made, they were held together simply by thongs, until, much later, primitive forms of the needle were devised. The countless instruments such as spinning wheels and looms that have developed from this primary impulse are common knowledge.

Of course, climate is an important influence in the development of fashion. In ancient Egypt, dress was wanted to be cool, simple, and comfortable. The wealthy women wore costly, veil-like draperies which were a protection against gnats and flies, but had no warmth and very little concealing quality. But while the Nile dwellers were still wearing the flowing garments that so little resemble modern dress, barbaric men in the northern forests had garments not at all unlike the coats, trousers, hats, shoes, and gloves of the present day. The idea represented in these garments was that of combining the greatest amount of freedom for the limbs with the maximum protection. This is achieved much the same now as then, so that one is struck with the relatively small departure that has been made, so far as essentials are concerned. In climates that varied from summer to winter it was logical to add more or heavier clothes in cold weather, so that winter and summer clothes were customary, though not at all universal.

Certain other derivations showing the influence of climate may be traced in modern clothes, however, that have lingered after the incentive for their origination has departed. One of these is the V-shaped nick on the coat lapel, which goes back to a time when collars could thereby be more easily turned up around the neck in inclement weather. This nick is seen on all men's coats of today, although so cut that to use them in this earlier way would be quite impractical.

But climate is not always logically followed. The Arabs, though living in a country that is extremely hot, are heavily clothed from head to foot, while way down at the very tip of South America, where they see icebergs every hour on the hour, the natives of Tierra del Fuego wear nothing but a skin that is hung from the neck by a cord and covers the front, back or side depending on which way the wind is blowing, its purpose less useful than decorative.

In nature it is the male birds and animals that are the most decorative, but in human life that has been reversed. This curious fact is due, however, not to instinct but to convenience. Among the early savage peoples the men, following their instincts, at first kept the wonderful shells and skins and other things that appealed to them for their own personal decoration. But when man had to go forth to battle for food or shelter, or to creep stealthily upon some animal, these things were a handicap in the natural carrying on of his work. They were heavy, or noisy, or in some way interfered with the attainment of his purpose. He therefore compelled his woman to wear them for him, to display his prowess, wealth, or importance. Whenever, through the ages, man has had the wealth and leisure, he has made himself as gorgeous as possible, and has often deliberately dressed in ways that made any work impossible, just to show his ability to pay for service. In our modern democracies this attitude has ceased to be considered a cause for pride, so that in very recent times the tendency toward gorgeousness has been moderated. So too the rather opposite habit of wearing somber black to show the loss of a dear one.

One of the oldest clothing customs is this one of wearing mourning for the deceased. But the motive was very different from ours, which is primarily an outward indication of bereavement. The original idea of the mourning dress was to act as a warning to the general public that persons so attired were soiled. It was believed that a dead body was unclean and that this uncleanness extended to things and persons that touch it. The extremes to which this idea was carried among most primitive peoples are amazing. That a corpse was so widely tabooed doubtless arose from the fact of the decomposition which in warm climates soon set in. The frequent experience in those unsanitary days, that where one has died another is likely to, may have contributed to this, though no scientific idea of infection existed. Ancient writings are full of instructions as to the purification of those who have been around the dead before they shall mingle again in society. Often the houses were burned, the possessions destroyed. The Chinese wore hemp, which was cheap, for in order to get rid of the taboo the mourning dress must be destroyed when the season of grief was over. It is a cause for thanks that since the real purpose of mourning clothes is therefore no longer existent, we have no reason for afflicting the public with our troubles, but can bear the burden of our personal griefs privately and without display.

In those very ancient days before radio or any distant intercommunication, the Egyptian made no effort to copy styles from others for the mere sake of using what was popular with the masses. In fact, since there was little of man's work from which inspiration might be obtained, nature was his guide, and the motifs of growing things about him were translated into his decorations. The lotus, the asp, the Nile, were typical sources. This close adherence to nature gave his work strength. He was not merely providing for the passing hour but for the spirits and the gods, and the idea of preservation was a dominating one.

Woman's dress in the land of the pyramids was marked by what we could call a shameless display and a unique brilliancy of color. The effect of the scanty garments in vivid tones accentuated rather than concealed the figure. The chief article of attire was the deep circular collar worn around the throat. It was composed of jewels, metals, enamel, or beads, according to the wealth of the person wearing it. Feminine dress consisted of a tight sleeveless robe, like a clinging skirt, of a texture that defined the figure. It reached to the ankles and extended a few inches above the waist. Bracelets, anklets, and large round earrings supplemented the costume.

Perhaps somebody may have told you that old one about the tramp who knocked at the door and said, "Lady, here is a button. Will you sew a coat on it?" Now, if that tramp had lived in the fourteenth century, his sense of values would have been all right, for at that time the button was a rarity, and because its use was uncommon, the whole style of clothing was different.

The first garments were mere wraps, without clasps, catches, buttons, or any other device to keep them together. The Roman toga and the Japanese kimono, generously overlapping or fastened with a sash, are typical answers to a buttonless world. Of course an early forerunner of the pin was necessity. That took its first shape in a natural thorn or a piece of bone sharpened at one end. Feathers with sharp points also served, and indeed, our word "pin" is a close relative of the Latin word for feather, *pinna* or *penna*. Our writing pens, formerly made of quills, also come from this word. Some people say the word "pin" comes from *spina*, which is the Latin for thorn. But whichever you believe, early pins were naturally large, coarse, and uneven as well as troublesome to get and clumsy to use. In fact, they were really not good enough to pin your faith to.

One of the first developments in pins was to twist the top into a head of some sort so that the pin would not go entirely through the goods. This could be done only after metal was employed, as bones, thorns, and feathers all snapped too easily. Metal pins and brooches, however, were a favorite in the sophisticated times of ancient Egypt, and even the so-called modern safety pin dates back four thousand years to those same clever people. After all, every self-respecting baby has a crying need for safety pins, and doubtless mothers, then as now, were glad to take the cry out of the need.

Many of these Egyptian pins and brooches were not merely useful, but of high decorative value, with intricate and beautiful designs. Crisping pins, which we may suppose to be curling pins, are mentioned 'way back in the book of Isaiah, showing that even then the ladies knew the value of curls. But in Europe, most people were still using ordinary wooden skewers as pins for the hair and other purposes even up to 1550. An endless variety of quaint old customs, stories and superstitions are linked with pins. They were much used for magical purposes, for revenge, for enchantments in all parts of the world, and were generally considered to be a protection against witches and evil spirits.

One old custom in Saxon lands was drinking to the pin. That was not a toast but a measure. Pins were fixed on the side of a wooden cup as a mark of quantity. Tankards had eight pins set at equal distances from top to bottom. They held two quarts, so there was half a pint between each pair of pins. To drink to the pin was an early measure of prohibition to suppress drunkenness.

Another relic of pins was a garment once called pinner. It was an apron covering the front of the dress, formerly pinned on, and hence called a pin-afore. Pins have been known to be a handy defense for a woman when accosted by a stranger. Samuel Pepys tells on himself that such an incident occurred when he wanted to make the acquaintance of a very pretty girl he saw in church. She, he says, "did out with a pin, which effectively cooled my ardor." Women used to have an abundance of pins distributed about their persons, for they wore many more garments than now, and had no other way of fastening them. In the cases of those tight veils and collars which almost looked like nun's habits, we can imagine the discomfort of pins which simply would stick into the flesh like stinging nettles.

Numerous games and amusements have been devised with pins, one of the oddest being a pastime of making pin-pricked pictures. The pins used, when this was a hobby in the last century, were of several thicknesses, the broad lines and heavy shadows being pricked on paper with thick pins, and the finer work being done with thin pins. Today we sometimes see patterns outlined in pinpricks. Pinpricking was used in one of the tragic episodes in the life of the unfortunate Marie Antoinette. During her imprisonment, wishing to communicate with some friends who were arranging an attempt to release her, she sent them a slip of thin white paper on which she had pricked a message to Comte de Rougeville, who was to carry her to safety. The message was intercepted, however, so that her ingenuity was fruitless.

When steel pins were first invented about the nineteenth century they were so expensive that only the wealthy could afford them. That is why we still say it is a sign of good luck to find a pin in the street. It became customary, when a woman was married, to give her a certain sum of money with which to buy these luxuries. The expression "pin money" has survived to this day, though it now applies to her spending money in general, as pins no longer require a special appropriation. As a matter of fact, pins have become so abundant that one hardly cares a pin about a pin, and it is said that not more than one pin in every hundred is worn out. All the other ninety-nine are lost, and where they go is one of the modern mysteries. If they were strewn along the streets in that proportion, there would, according to the old saying, be a great deal of luck to go around for the picking.

Today pins are so finely made by machine that some varieties for special purposes can be had so small that 4,500 of them weigh but an ounce. The machinery by which pins are offered for sale in parallel rows on paper strips is almost as intriguing as the pin itself, and brings us a long distance from the time when one had to apply bow and arrow to a bird on the wing, or tear one's hands among the briars to get a pin from nature. Thus the pin has pricked its way through the fabric of our lives, giving accent, sometimes painfully, to this existence. Pinpricks are recognized as troublesome, persistent little ills. Life is full of pinpricks, and it seems an excellent way of expressing those small but exasperating worries with which we are so frequently annoyed, yet which do not matter so much after all. In most cases we endure the discomfort of the pinpricks for the sake of the comfort of the pin—and when we want a pin, we want one!

The use of the button as an article of dress is a product of modern civilization. Were it not for buttons, we might all be wearing flowing robes, for modern dress would be difficult to manage without them, especially in the absence of hooks, elastic, snappers, zippers, and the other fasteners we can now resort to. So if men go down to the office dressed in tweed suits instead of Roman togas, it is because someone invented the reliable button.

Buttons, though known in ancient times, were used merely as ornaments up to the fourteenth century. This is one of the peculiar instances in which the general custom of making something useful and then decorating it is reversed. Here someone discovered an object apparently only decorative, and not until much later was practical use found for it. The word "button" is of French origin, and meant a knob or something that stuck out or butted out. Fourteenth-century pictures show buttons on fancy clothes, but no buttonholes. They were often rosettes of cloth, sometimes of gold and silver, all of course made by hand. They were often elaborately engraved and were very expensive. It is from this time that buttons became a term of comparison for something trivial and insignificant, before their potential value was realized. Rows of them would be found from throat to waistline, or running all the way down the arm from neck to wrist, all apropos of nothing. The next step was the use of loops with buttons, which made them practical; and the buttonhole came last of all. While the decorative purpose of buttons has continued, they began to be employed as fastenings about the fifteenth century.

Even a so-called trivial button has been known to do significant work. But for a button, the world would never have thrilled to the majesty of the *Messiah.* The button in question was only an ornament, yet without it, we would never have felt the awe and rapture of the exultant, inspirational Hallelujah Chorus. For a small button was the shield that saved the life of youthful George Frederick Handel in a duel with a fellow musician. Struck by the point of his adversary's sword, it spared the life of the composer and his glorious music for the ages.

In America, the button is a sort of first aid to celebrators, and almost every organization that has a convention or an anniversary marks the occasion with a button. Buttons proclaim your political party at election time. They tell whether you are a Mason, a Boy Scout, a college senior, or a bellboy, and whether you have contributed to the Red Cross. On a uniform they tell whether you are a fireman or a policeman, while military buttons often announce the rank of the wearer. Buttons now are made of every material—metals from mines, bones from the stockyards, the palm trees of South America, the oyster shells from the bottom of the sea, as well as porcelain, leather, plastics, and glass. The buttons which we find on our commonest clothes may have traveled more than we ourselves in order to serve us.

Some buttons still appearing on men's clothes are not of practical use, their original purpose having been forgotten. The pair of buttons found above the tails of a full dress coat, for example, are those which once held in place the sword belt that completed the costume of the well-dressed man. Sword and belt have gone from all but military dress, but the patient tailor still sews the buttons on the same spot. The origin of buttons on the sleeve is attributed to the fact that Frederick the Great was very particular about the appearance of his soldiers and their uniforms. He discovered soon after taking charge of his army that many of the men were in the habit of wiping the perspiration from their faces with their coat sleeves. This of course tended to soil the sleeves and give the coat an untidy appearance. In order to put a stop to the practice, he ordered that a row of buttons be placed on the upper side of each sleeve. After that, when a soldier undertook to draw his sleeve across his face, the buttons would make it so uncomfortable for him that he soon learned to abandon the habit. By this ingenious means the untidy practice was broken; but though the habit disappeared, the buttons remained, and when their original purpose was forgotten they were placed on the under side of the sleeve to be out of the way.

This makes us wonder where the humble handkerchief was hiding. Today a handkerchief is more than a charming accessory of dress. There is something intimate, something so peculiarly personal about it, that it has always been idealized. Perhaps the daintiness of a lingering perfume is often accountable. But whatever it is, certainly handker-

'he Jester, an engraving by E. Mohn
ıfter a painting by A. Lambron. His
›uttons were purely decorative.

N.Y. Public Library

Officer of a Prussian cavalry regiment (1740)

chiefs more than anything else have been chosen by sentimental lovers to steal and treasure as precious mementos of their loves. Efficiency and practicality took a long step forward when these succumbed to the popular unsentimental paper variety.

There was of course a time before handkerchiefs, as we know them today, were thought of. Before handkerchiefs came kerchiefs. This word was a combination of *couvrir,* to cover, and *chef,* meaning head. Savage races wove grasses into small mats, and both wore them on the head and used them to wipe the sweat from their brows. From so modest a beginning did we derive both hats and handkerchiefs. The Chinese are the first known to have used them. With them it was common in its primitive silk tissue and paper form long before the Christian era. It then wandered to the land of the lotus, where it was held to have talismanic properties. In ancient Greece and Rome handkerchiefs were known, and were simple squares of flax usually tucked beneath a girdle.

Churchmen were the first to use handkerchiefs generally in Europe, and for a long time priests alone were permitted to carry them. In spite of occasional references to handkerchiefs, they were extremely rare throughout the sixteenth century. The queen of Louis XI had but three, and King Henry IV had only five. That they were once highly prized is shown by their being mentioned in wills. Thus in 1562 one was willed to a near relative.

When noses required attention, I regret to say that people applied either the cuffs or the fingers. It is interesting to know that at that time forks had not yet been invented, and people ate out of a common dish with the fingers. Therefore, good breeding demanded that persons of refinement should touch their food only with the right hand and their noses with the left. Indeed, from the earliest times the left hand was always reserved for unclean purposes, and most religions consider the left-hand side unlucky. Bad omens came from the left, the word for which was "sinister," and guests of honor are placed at the right of their hosts as a mark of distinction.

Beginning about the sixteenth century, silk handkerchiefs embroidered and fringed, or laced with gold, were known in England under Queen Elizabeth. The development of the handkerchief about this time became rapid, and in France particularly it reached an elegance scarcely conceivable. In the seventeenth century it was made of exquisite hand-wrought laces and was sometimes even ornamented with gems.

Do you know why handkerchiefs are usually square? It is because this shape was fixed by royal decree, and usage has perpetuated the form. At one time handkerchiefs were of any shape that individual fancy dictated—round, oblong, oval, or what not. One day at Trianon, Marie Antoinette happened to mention to Louis XVI that she was tired of these various shapes. The king at once decreed that "The length of handkerchiefs shall equal their width throughout the kingdom," thereby settling the matter for the years to come.

The introduction of snuff into polite English society under the Georges brought the handkerchief into very prominent use, and it became an art to handle it with elegance and grace under the stress of trying to conceal a sneeze.

Among the French, formerly, to make even the most casual reference to the handkerchief was considered the height of vulgarity. But woman's coquetry and daintiness removed this prejudice, if such it may be called. To the Empress Josephine we are indebted for reintroducing the handkerchief to court, which she did with a dainty *mouchoir* that had a wide lace border. This she would often raise to her lips when smiling, in an effort to conceal imperfect teeth.

Before our savage ancestors dreamed of the handkerchief or even had clothes to wear, they were concerned with methods of effectively dressing the hair. To them, hairdressing was usually decorative, devised to increase personal beauty and to lure the other sex. Often, however, it also symbolized a certain age or rank in the community. Early Egyptian headdresses were extremely elaborate, among both men and women. They were cut sphinxlike in square lines and fell like a curtain in back to afford protection to the nape of the neck. They really took the place of hats. The hair was divided into many tresses, each thickly plaited and curled, crimped, and massed in various ways. Wigs were already popular. The Greeks and Romans also dressed their hair most painstakingly, although in the classic age their methods were simplified at least to reveal the contour of the head. Men adopted the close-cropped manner of today. Razors are prehistoric—not the steel safety type of course, but of stone, flint, obsidian, and other cutting materials. It was left for woman to retain her crowning glory.

This was by no means permanent, however, for throughout the Dark Ages men permitted their hair to grow long again, and this fashion continued until almost the nineteenth century. From the time of the Crusades they began to give as much care to their flowing tresses as did women. England today, ever loyal to her past, has her barristers outfitted with wigs in deference to this former style. In the early Middle Ages the wearing of long hair was a sign of liberty, for serfs' hair was cut. Louis XIV, on account of the excrescences on his head, adopted a high wig, and the whole of Europe wore wigs until the French Revolution.

Fashions in hairdressing seem to vary as much as differences in clothes, and all sorts of peculiar and fantastic methods have seen their day. During the early part of the reign of Louis XV, when the huge pannier dominated costume, hairdressing was so simple that the head seemed disproportionately small. As the gown decreased, the coiffure became enlarged, and by 1750 it was a great edifice supported on a wire frame, stiffened with haircloth, and surmounted with a small beflowered cap or hat that nowhere touched the head but stood upon the foundation of curls.

In England at the same time the extreme was carried still farther. There are prints of headdressings so tall that the bearer had to bend low to enter a doorway, and others that required the hairdresser to mount a stepladder to reach the summit of a coiffure that he was attempting to finish. The numerous recipes that the literature of the times contained for the destruction of insects give a sidelight on the possible state of the lady's head when her elaborately designed hairdressing was perhaps once a month taken apart and renewed!

Elaborate and artificial coiffures were not limited to the lady. A gentleman of the eighteenth century would have felt as ridiculous without his wig as one of today would feel without a collar. The custom of powdering the wig is said to have originated through the whims of some French buffoons. A troupe of these performers, wishing to make themselves as grotesque as possible, covered their wigs with flour. This caught the fancy of some French men about town, who imitated them, and soon established a custom which came to be regarded in all seriousness. At a time when thousands of human beings were starving in France and England, immense quantities of flour were wasted on hair powder. Delicately scented powders soon replaced ordinary white flour, and great powdered and perfumed headpieces, costing sometimes three hundred dollars or more, came to be part of the dress of every well-groomed gentleman.

These extravagances gradually abated for both sexes, and in the first part of the nineteenth century women were dressing their hair very simply, letting it fall at the sides in ringlets and tying it at the back with a ribbon. Then somewhat later the side curls were dispensed with and the hair was arranged on top of the head. Often it was drawn over a cushion or other support to make a large roll, an arrangement known as the chignon.

Hair bobs are not so recent as many of us have believed.

At two or three periods during the seventeenth and eighteenth centuries, both in England and France, this style was a fad. Now it seems that the bobbed head is an established custom, although the fashions occasionally oscillate between short and long. While long hair symbolizes the strength of Samson, short hair symbolizes woman's recent independence. But symbol or no symbol, hairdressing today is so much more comfortable and convenient than in the complicated manners of the past that it is as charmingly simple as it has become simply charming.

Vying with hairdressing in the matter of fashions were the shoes plodding through the ages. Many sidelights on the past are revealed in the story of the shoe, which is by no means the conventional or prosaic utility that we ordinarily think of when this word is mentioned. The first reference to footwear found in writing is said to be a sacred Hindu composition known to be thousands of years old. The inscription directs all worshipers to remove their shoes during devotions at certain shrines. Thus even at that early period it would appear that shoes had already entered into the tradition of the people. "Loose thy shoe from off thy foot, for the place whereon thou standest is holy." Thus spake Joshua. Loosing the shoe is a mark of respect in the East even today. The Moslem leaves his slippers at the door of the mosque. He comes barefoot into the presence of his superiors. This custom is a mark of humility; since the shoe is a sign of dignity, the shoeless foot is a mark of servitude.

Of course, the early shoe was a very simple affair, consisting only of a sole tied to the foot by a thong passed between the toes. The sandal indicated class or social distinction in Egypt. Slaves and captives were obliged to go barefoot or to wear a very primitive sandal of woven palm leaves. The "common people" were permitted to wear sandals of any material, but they could not have long, pointed toes, these being reserved for victorious warriors or other persons of high rank. Color also indicated caste.

Red and yellow were the favorites of the highest rank, and there was trouble ahead for others who affected them.

The small feet that for so long the Chinese women cultivated were not supposed to have esthetic beauty, but only to indicate the woman's faithfulness, symbolizing the fact that she could not rove far from home. The swaying, hesitating gait of the Chinese woman was a result of maltreating the feet. Again, the pigeon-toed gait of a Japanese woman was attributed to the closely clinging kimonos constantly pulling the right foot inward. Japanese went barefoot or sometimes wore *zori*, or sandals. They also had *geta* or wooden clogs that were three inches high, and the only way one can keep them on and walk in them is to bend over slightly to maintain the balance. This gave the concave appearance of so many Japanese. Since in the Orient people of low stature were rather disdained, high heels came into existence to increase apparent height. One can easily imagine that it would be annoying to have the clog on one foot make a different sound from that on the other. The Japanese, in selecting their clogs, are careful that the sound emanating from each shall be alike.

Grecian women of high rank sometimes had as many as twenty-two kinds of shoes. When they left their houses, slaves accompanied them, carrying a rich carpet in which were placed shoes appropriate to the occasion and the place where they were going.

Since ancient days shoes and weddings have been associated in popular folklore. The familiar custom of throwing old shoes after a bride is but one of the ancient ways by which a father transferred his authority over the girl to her husband, the shoe being the symbol of authority. At an Anglo-Saxon wedding the shoe was as indispensable as the ring today. During the ceremony the bride passed her shoe to the groom, who signified his future authority by tapping her on the head with it. In the Tyrol it was once the custom for the groom to present his bride with the wedding shoes. These were covered with dainty heart-shaped decorations, and as long as this token lasted, it was believed, her happiness would endure. The idea of the shoe as a symbol of possession survives in many of our everyday expressions, such as saying that a person has stepped into another man's shoes, when we mean that he has assumed the authority that belonged to this other person.

The Crusades, as much as anything, were responsible for the beginning of modern shoes, for during those long pilgrimages durable shoes became a necessity. But the dandies of the land, who did not work and were proud of the fact, went in for highly impractical forms. The fashion of wearing long, pointed toes came originally from France, where they are said to have been devised by a Count of Anjou to conceal his deformed feet. As time went on, the length of this pointed toe so increased that it became necessary to stuff it to prevent its doubling under the foot and being tripped over. By the fourteenth century such shoes had become even more freakish. The points were so long that they had to be fastened just below the knee by a garter. Very elaborate fastenings, often of silver and gold, were used. The clergy objected to this style of shoe, for the reason that those wearing them could not kneel in church to say their prayers, and then the state took up the matter. Finally a proclamation was issued which limited the points to a length of two inches.

During the middle of the sixteenth century the women of the French court wore such high heels that the most fashionable ladies were literally standing on their toes. Once in Venice it was the fashion for women to wear shoes so high that it was equivalent to walking on stilts. A wooden platform the size of the foot was built, and the shoe perched on top of it. The mode was introduced by certain Venetian nobles to discourage their wives from gadding about. They made it difficult but not impossible. No diminution in "gadding" is reported!

In the reign of Henry VIII of England, this gentleman

had such swellings on his feet, caused by gout, that shoes of very broad design were made the fashion. During the reign of his daughter Mary, shoes had grown to such size that Parliament was compelled to limit the breadth over the toes to six inches. A high heel became especially popular in the fashion set by Louis XIV of France. Being very short himself, he wore these high heels to raise himself some inches above the ground and thus look more imposing.

The word "pump," which is now applied to a light kind of thin-soled shoe used especially for dancing, probably has its name from being first used for ceremony or "pomp." It was almost the nineteenth century before right and left lasts were widely known for shoes. Until then they had been such "quick fits" that they might be worn on either foot with equal discomfort!

Only soft leather shoes even pretended to be comfortable until the end of the sixteenth century, when woven stockings were introduced. Before that, stockings had been tailored from pieces of hard fabric, the seams of which precluded the use of any but the softer kind of shoes. But with the woven stocking came the snug-fitting shoe which was the forerunner of the conventional footwear of the present. History tells us that stockings of a crude cloth or leather were first made about the seventh century in France. The Scots produced the first knitted hose toward the end of the fifteenth century, and these were made of wool. Before that the stockings worn by those who wore them at all were of linen, cloth, blanket cloth, or velvet, according to the wearer's station in life. But these were not stockings in the sense that we know them, and the word "stockings" came in with the knitted article. Queen Elizabeth was presented with the first pair of silk stockings and liked them so much that she completely lost interest in any other kind.

About sixty years ago there was a young man who had a job in a factory in Lowell, Massachusetts. In this factory, as in so many where heavy machinery was in use, the building was constantly jarred by vibration. Not being very robust, he found this vibration very disturbing and pondered over how he could relieve or at least diminish it.

One day be brought a rubber mat to the factory and stood upon that. Immediately he found that the rubber absorbed a great deal of the movement. Unfortunately, a few days later the rubber mat disappeared, so our young friend had to think of something else. He did. He got two little pieces of rubber and nailed them on his heels. His idea gave him two little rubber mats that could not be stolen by anyone. The name of the young man was Humphrey O'Sullivan, and this was the origin of rubber heels, of which he was the inventor. Today his firm is one of the largest manufacturers of rubber heels in the world, and men and women are walking about their business in a resilient, quiet, easy way not known before his improvement.

Clothing, of course, depends upon textiles, and recent years have radically transformed this area of human interest. Today we no longer rely upon the four basic textiles —flax, cotton, silk, and wool. Now man manufactures textiles synthetically and according to prearranged specifications. They are now available crush-proof, unshrinkable, nonstretching, fire-resistant, stain-proof, unrusting, color-fast, and a whole assortment of other remarkable qualities. Man's ingenuity has utilized the strangest materials for his magic —coal, cheese, glass, and many others. The results have extensive effects on business and industry, and enhance beyond measure the convenience and beauty of today's wearing apparel.

X

The Long and Short of Dress

THE PRIMAL URGES of man were three—namely, for food, shelter, and clothing. The satisfying of these brought, in the long run, the gain of power and the acquisition of wealth, which caused him to develop a fourth urge, the love of distinction. As the masses gained in freedom and education, knowledge brought them increasing acquaintance with things of beauty and a longing for their possession. The desire for distinction is born of the wish to rise above the commonplace, to give expression to the individuality of which the human soul is in varying degrees conscious.

This sense of distinction has been called the keynote of modern fashion. It is an outgrowth of the pride of personal accomplishment, though often the one with an inferiority complex affects it to conceal his deficiencies. But finding the perfect style is not keeping it. A fashion may perhaps be all that is beautiful, efficient, durable, becoming, yet have only a temporary existence. For as in all life there is perpetual change, so here. The desire for variety is merely an indication of the ever-expanding inner life of the human being, the constant adjustment to new conditions. Surely no one would venture into a New York subway attired in a hoop skirt. It seems regrettable that the modern man's attire is as conservative and uninteresting as it was a century ago. It has neither comfort nor picturesqueness to recommend it, and seems to offer a fertile field for mother wit to focus on.

"Style" is the great word in the vocabulary of women's clothes today. Every representative shop has its stylists—

those who anticipate the trends and provide accordingly. In some ways it seems like a terrific economic waste, though to so many people it gives an interest in life and leads to a superficial sort of prosperity at least. The main objection to style is that people feel so absolutely bound to follow it. We do not keep our own independence and individuality. We alter things we like into things we do not like, or buy new ones while the old are serviceable. Shelter, food, and clothing. How simple could be the connotations of each of these, yet how elaborate have they become. It is doubtful if they have not been overstressed and exaggerated. They add to comfort, efficiency, and beauty only occasionally, and as often retrogress in these directions as not. Many people assume that when persons are well clothed, well fed, and well housed, they will go on from there to better cultural pursuits. However, instead of turning their attention to intellectual or artistic interests, they often tend to exaggerate their desires for still more expensive dress, food, and shelter. Most of us have but a total of twenty-four hours a day to utilize; and as we give these hours to such matters out of proportion to their importance, it is obvious that we must take them from considerations that might be valued more.

Meanwhile the merchant has found new romance in his business. He is no longer just a seller of fixed commodities. His office is a channel for world thought; past him sweeps the kaleidoscopic panorama of modern trends and ancient customs. He watches political and economic movements. He studies the art of his own and foreign countries. A great merchant tells us that out of these, styles and fashions grow.

The dress of nations has resulted, we have seen, from a variety of influences—climate, occupations, history. Wars and religious movements, patriotic crises, have left their impress. Nowadays, regrettably, garments reflecting tradition and embracing native crafts are passing. Today traditions are forgotten almost before they are established. Only in primitive lands and remote provinces do we see vestiges of more picturesque days. The styles now generally worn by men and women of all classes are dictated for the most part by fashion and convention, more or less arbitrary, rather than as a matter reminiscent of racial pride and glory.

The costumes even of Holland are rapidly passing away. But if these costumes have picturesqueness to commend them, at least they are not comfortable. The wooden *klompen* are noisy and heavy; to hear school let out and all the children running away in their wooden shoon is to think of nothing less than a regiment of horses and drums and a burst of thunder. Though I visited Holland in hot weather, all the women wore quantities of heavy woolen petticoats of various colors, and tight bodices. So too the little children, and it is entirely obvious that they wear corsets too. I am referring to such suburban regions as Marken and Volendam, where native costumes survive, though the lovely lace caps, so charming and graceful, do sometimes wander into the larger cities. The men in these quaint places tie scarves about their necks, and their shirts are generally patched in some extraneous material and color, though I don't know if that is part of the costume or not. Their trousers are skimpily short and enormously baggy at the top. With pockets such as his, no Dutchman could ever abstract anything so small, say, as a nickel, and get into a "pay as you enter" car the same day. Still, this is one of the many things he is spared, for all his habits are based on the leisurely attitude, and to see a Dutchman in a hurry is to discover how slow a hurry can be. Even the few remaining brown, weather-beaten windmills lazily flap their big arms as if there were plenty of time in which to do the day's work and no need of rushing it. Because of the bleak winters in Holland, when even the canals are frozen over and one uses skates instead of boats, heavy gloves are added to the costume.

In fact, the practice of protecting the hands with some sort of covering is a very ancient custom. Whether against cold, or against rough usage in passing through thickets, the adoption of some kind of cover, rudely fashioned from skins of animals as it might be, was to be expected. In warfare and the realm of sport, gloves were at the earliest times employed as both symbols of defiance and weapons of offense. They consisted usually of thongs of leather wound around the hands and weighted with metal. Often a single blow with one of these would kill a man. They were first fingerless, like mittens. The Romans in their warm country regarded them with more or less contempt.

According to old Germanic law, the hand was the great symbol of power, and no doubt the glove derived some of its significance from that fact. In the Middle Ages the glove enjoyed great importance as a sign or title of investiture. The Earl of Flanders in 1294 delivered the towns of Bruges and Ghent to Philip the Fair by handing him a pair of gloves. At one time no town of Saxony could establish a free market without imperial sanction, and in token of his permission the emperor was wont to send his right-hand glove. Its convenience as a loose object made it a handy missive. In the days of chivalry, for a knight to cast down his glove or gauntlet was tantamount to a challenge to combat, and the usual mode of acceptance was for the challenged to take up the glove, at the same time casting down his own.

The wearing of gloves was originally restricted to the ranks of clergy and soldiery, again to symbolize their class distinction, and their use outside the Church or army was interdicted by law. The custom of wearing gloves more generally hardly came into being before the time of Henry I. It is probable that the scope which the glove offered for ornamentation was the first impetus to its general adoption. There are evidences that early in its history it developed claims to be deemed not only a utility but an article of fashion as well. Embroidered gloves and gloves with bejeweled backs were in vogue in the twelfth century. The foppery of young Norman nobles came in for considerable criticism, and at least one writer of the time is on record for saying quite bitterly that they covered their hands with gloves too long and too wide for doing anything useful. Thus the glove was still a symbol of class consciousness. It was not until the thirteenth century that women wore them, when they were considered ornaments, and were of linen, extending to the elbow. Queen Elizabeth had them in greatest elaboration.

The French have at least two phrases to show the emblematic character of gloves: *avoir les gants,* to possess the gloves of, or to have the priority in, and *se donner les gants,* to give one's self gloves, in the sense that we say to give ourselves airs.

In the early nineteenth century, gloves were so seriously regarded as accessories of fashion that one man wrote a series of rules about them in which he describes six pairs of gloves which an English gentleman should wear every day for different social purposes: to drive to the hunt; actually while hunting; to return to London after a drive; for a walk or to visit a lady; to go to a dinner party; and for an evening ball.

Today gloves have lost much of their symbolism, but they are an accessory of refinement and good taste, and to that extent they bespeak some of their former distinction. Since, however, they are within reach of everyone, they do not offend our democratic ideals, and of course today they may be so carefully fitted that they afford a comfort and grace to the wearer that were not always expected.

The symbolic history of gloves somewhat parallels that of hats. When a gentleman raises his hat as an act of courtesy, he is but continuing a custom that had its origin in the days of knighthood. Whenever one of our chivalrous ancestors entered a crowd or gathering, he was careful to have on his armor and helmet to protect himself as far as possible from foes that might be waiting an opportunity to dispatch him. But when coming among friends, especially if they were women, he removed his helmet to indicate that he trusted them. The modern hat, as the successor of the helmet, no longer serves as a protection from enemies, but is still used to show respect and courtesy. Not so long ago hats were made in only a few sizes. Inside, however, was a lining on a drawstring so that they could be taken in or let out to fit various head sizes. The little bow inside a man's hat is a relic of this kind of millinery. The word "millinery," by the way, goes back to the city of Milan, where straw for hats is still one of her desirable products.

But older than the hat as a protection for the head was the sacred umbrella. We are so apt to think that all mechanical devices belong to our own day and age that it is something of a surprise to learn that umbrellas are in no sense a modern invention. In fact, they are found sculptured on the monuments of Egypt and the ruins of Nineveh, and their use in China and India also dates to antiquity.

The word we use for this object is derived from the Latin *umbra*, which means shade. The umbrella was primarily a sunshade, its original home having been in hot, brilliant climates. "To take umbrage" is a familiar expression derived from the same word, and in this context means to take a shady or gloomy view.

From the earliest times in Eastern countries the umbrella was one of the insignia of royalty and power. Pictures and carvings have come down to us showing kings in procession with an umbrella being carried over their heads. Certain princes of India had among their titles that of "lord of the umbrella." In 1855, while addressing the Governor-General, the King of Burma referred to himself as "the monarch who reigns over the great umbrella-wearing chiefs of the Eastern countries." Certain sizes and colors are used by kings and princes, while smaller sizes and other colors are used by the lesser nobility in some oriental lands. The baldachins erected over ecclesiastical chairs, altars, and portals, and the canopies of thrones and pulpits, are in their origin closely related to umbrellas and have the same symbolic significance. In each of the basilican churches of Rome there still hangs a large umbrella.

There is also another expression, "under someone's umbrella," meaning under his dominion or influence. This alludes to the umbrella as an emblem of sovereignty. Not so long ago it was written that the retinue of the Sultan of Morocco was composed of a troop of from fifteen to twenty men on horseback. About one hundred steps behind them came the sultan, mounted on a mule, with an officer bearing his umbrella, who rode beside him on another mule. Nobody but the sultan himself, not even his sons and brothers, dares to make use of it. Those who have ever seen *The Mikado* of Gilbert and Sullivan will recollect that that dignitary also had his special officer of the umbrella.

In Greece, umbrellas had a part in certain religious ceremonies, and it is evident from paintings on ancient Greek vases that umbrellas very much like ours were known long before the Christian era. In Rome also they were used, but only by women, and could be closed when not in use.

In the sixteenth century the use of umbrellas became very general in Italy and France. In fact, they soon became articles of luxury and fashion. Their color, material, and ornamentation indicated the rank and wealth of their owners.

The use of umbrellas in England, in spite of the rainy weather, was not extensive in early days. Allusions are found to them in ancient manuscripts, but an English dictionary published at the end of the seventeenth century still defines it as "a screen used by women to keep off rain." A man named Jonas Hanway is said to have been the first to carry an umbrella through the streets of London in rainy weather, about 1750, and he was hooted and jeered at by the boys for his fears of a wetting. Incidentally, he created quite a disturbance among the sedan porters and the public coachmen, who, like the modern taxi-drivers, resented anything that would deprive them of a likely fare.

The umbrellas used in Hanway's day were clumsy, ungainly affairs, ugly in color and not altogether rainproof. They had long handles with a crook at the end, and the ribs were thick whalebone, while the cross pieces were of wood. They were loosely jointed and rough. The covering usually consisted of oiled silk or cotton, very heavy, and was apt to stick together in the folds. One must have weighed ten pounds when water-soaked.

Umbrellas were introduced in America toward the end of the eighteenth century. A Baltimore man bought an umbrella from a ship that had just returned from a trip to India, where the umbrella was used as a sunshade. On his way home with his purchase, the idea of using the umbrella as a rain shield suggested itself to him. At the first sign of rain he paraded with his new weather shield down the street. At the sight of it women ran from fright; horses became wild. The children were braver. They banded together in groups and followed the innovator, staring in wonderment. Soon he was threatened with fists for disturbing the peace of the town. He was mobbed and his umbrella torn to shreds.

The news of this episode spread far and wide. To Philadelphians, who were more liberal with innovations, the idea of a rain shield such as this was very attractive. Before long it was a common sight there, and soon New York and other parts of the country were having them. In some villages it was considered a mark of distinction to be seen with one.

Now the use of an umbrella is permitted to everybody. Not only may the king be sheltered from rain or sun, but the simplest of us may have one. Umbrellas are frequently made to double up into traveling satchels. They are also made in colors to supplement milady's various costumes. Their handles are works of art, and their mechanical aspects partake of the flavor of our efficient twentieth century. Men carry them without a quiver, and even little children may be provided with those in small sizes that they can readily manage.

Many interesting oddities of costume that have come to us are so sketchy that it is difficult to group them under one heading. For instance, Sir Walter Raleigh wore a corset, as did fashionable noblemen for many decades; even Lord Tennyson refers to his creaking stays. In questions of size, the higher the rank the greater the extremes permitted. The queen always had the longest train, the widest hoops. Monstrous ruffs were introduced by Queen Elizabeth in order, it is said, to conceal a somewhat yellowed neck. Spoons had long handles, often with a bend in them, to give access to the lips over these stiff starched collars. All these fads were determined by fashion's decree, which in turn was the result of factors that we would not normally connect with them.

Fashions of old-time status seekers often blossomed out in impractical forms to indicate that the wearer was of a social position not requiring him or her to work. Such was the reason, originally, for the very high, thin heels, the tight waists, the narrow skirts, of caste-conscious Europe, for the long, long nails of the mandarins, and many other follies.

In some countries changes were effected in fashion by the plagues that swept over them. London in 1665 was almost depopulated, and it was believed that imported garments and materials were a cause of infection. Even the fashionables became chary of buying new clothes. Tailors were accordingly compelled to close up, and men who were ordinarily well dressed wore the shabbiest things. When they did buy, it was very sparingly, so that simpler things came into use. In smaller measure, the wearing of wigs was also affected by plagues. Many of the wigmakers' products were made from hair clipped from the heads of plague victims, but human vanity was such that even risking death itself was thought preferable to exposing gray hairs, or no hairs at all. Men could stand shabby garments better than the signs of age. So the wigmakers flourished while the tailors starved.

Occasionally a monarch attempted to curb extravagances in headdress and clothes in general. Charles II himself adopted a plain type of dress similar to our modern suit, in such an effort. He declared that he would wear no other style during the rest of his life, and despite the secret smiles of his courtiers, he kept his word for quite a while. But his luxurious neighbor, Louis XIV, hearing of this decision, promptly adopted the English monarch's costume as livery for his servants. This was more than the reformer could tolerate, so Charles returned to his earlier modes.

The modern type of man's dress, dark, well fitting, and standardized, is due to Beau Brummell, who initiated clothing with these characteristics that have continued in modified form to the present. Brummell had set up a bachelor establishment in Mayfair, and thanks to his friendship with the Prince of Wales (later George IV) and his own taste, he became the unquestioned leader in men's styles. His social success was instant and complete, his repartee the talk of the town, and he had a remarkable talent for presenting the most ordinary circumstances in an amusing light. But eventually gambling and extravagance exhausted his fortune (he had inherited about thirty thousand pounds), while his tongue proved too sharp for his royal patron. They quarreled. His popularity declined. He fled to Calais to escape his creditors. He lost all interest in dress, his personal appearance became slovenly and dirty, and he finally died a wreck in a charitable institution in France.

No Beau Brummell but custom alone developed styles in India. There the most typical and beautiful of the feminine garments is the *sari*. This voluminous drapery is a long, straight piece of silk or satin wondrously colored and exquisitely embroidered. It is wound round and round the body, outlining it tightly at the back and hanging in straight folds in front, each evolution being accomplished in pleats to attain sufficient fullness. Finally the end is drawn over the head, concealing one ear and leaving the other exposed. The article of masculine attire to which most importance attaches is the turban, its shape and general aspect denoting the social status of its wearer.

Dress to the Persian is a very important matter. While in the Western world rank is denoted by a coat of arms, in

Persia an aigrette is used, fastened to a turban, or a feather in a hat may serve the purpose. Where turbans came from no one really knows; maybe from Assyria, or perhaps from India. But in Persia, whether or not you are in style depends upon the way you wind your turban. Any turban consists of between ten and twenty yards of cloth wound over a cap like a Turkish fez.

In China we find a people who for some reason are rational, strictly considering utility in a way that is less familiar here. They early evolved a rational dress which is not exactly beautiful, as it is neither gravitational nor anatomic, neither following the body nor the fall of texture, but avoiding both extremes; and having evolved it for use, they changed it little.

All over China, and particularly in official circles, dress was determined by certain fixed laws, the result being that every detail possessed a meaning for those capable of interpreting it. The most significant feature was the button which adorned the crown of the hat peculiar to mandarins, while embroideries likewise assisted in determining the status of the wearer. Color was another factor of importance, yellow having been sacred to the emperor and his family; red was exclusively reserved for mandarins, but blue, violet, and black were common property. After a person reached the age of eighty, no matter what his rank, he might select his wardrobe from any colors. Another distinctive feature of a mandarin's dress was the pectoral, a small piece of material attached to the breast. In the case of civil dignitaries it was embroidered with birds, while in that of military authorities it displayed quadrupeds.

Color to the Chinese is not a matter of chance nor merely a matter of pleasing the eye; each color has its own definite meaning. Thus yellow expresses the thought of grandeur, power, and riches. Green symbolizes everlasting peace. Red is the color of happiness as well as longevity. White is the symbol of light and purity and also indicates mourning.

The flowers of the East are an important inspiration to

her artists. Chrysanthemum, iris, hydrangea, carnation, wistaria, water lily, plum tree, bamboo, cherry blossom, twine their leaves and branches through all the artwork of the East. Then there are the animal forms that play so conspicuous a part on all things Chinese. The dragon is the spirit of fertility. The pheasant was the emblem of the empress and is said to protect women. The cock is the emblem of the sun. The horse is the symbol of wisdom. The dog foretells good fortune. The crane denotes long life. If the Chinese lady is the wife of a mandarin of high rank, she has the symbols of her husband's status embroidered on the long surtout covering the dress.

The Chinese woman keeps her gown hermetically closed up to the neck, with a straight fastening, rigid as an officer's collar, while the Japanese wears the shawl collar, leaving the neck free. Whatever her condition, the Chinese never wears a low-necked dress or allows the least flesh to be seen on arms or legs. The Chinese dress is never tight fitting. What makes it beautiful is its embroideries, the fineness and variety of the materials that vary from a sort of voile or crepe to thick satin, but chiefly the happy assembly and harmony of colors. Chinese and Japanese art is known for its attractive brushwork. Because writing in these countries was accomplished by means of a brush instead of a pen, artists were trained from boyhood to use the brush with great delicacy.

In Japan, as in other oriental countries, the people worked in their native provinces and produced their own kind of silk, linen, leather, or what not. Each province had its crest or seal, which was known as *mons* and was used on every article made in that province. Five of these designs were placed on a kimono, one on each side in front, one on each shoulder, and one in the middle of the back. The coolie laborers were an exception to this, however, for they wore only one, and that in the middle of the back. Sometimes this was on the clothing, but often it was on a large card suspended by a cord about the neck, for the upper part of

Pan American

The sand gardens at Kyoto, Japan, with girls in the traditional kimono.

the coolie's body was often nude. These *mons* or crests are often combined with other forms of decoration and are very interesting in design and color. They were originally used as family or state marks similar to the heraldic designs so popular in the Occident.

The pet vanity of the Japanese was to arrange their undergarments so that the border of each formed a regular trimming at the neck. A glance sufficed to show how many were worn, the greater the number the greater the success achieved by the wearer. At least two or three kimonos are placed one over another in this way, their edges showing about the neck. In bad or cold weather, extra ones are added for protection. The obi is a sash more than three yards long and about twelve inches wide. It is wound two or three times about the figure well above the natural waistline and tied in a great bow at the back. The brilliant colors of the obi vie with those of the kimono; and rich materials, such as gold brocade and others, make this article of attire often very expensive. In Japan it was the custom of dancing girls to wear gowns of bright hues, bright reds, yellows, or greens embroidered all over with gold or fancy silks. This being the case, the best families did not dress their daughters in bright colors after they were five years old. Older Japanese girls might wear a delicately colored kimono with a bright-colored sash or an embroidered band, but a married woman, especially one with a family, would wear only somber-colored kimonos with bits of color about the neck or sleeves. Sleeves of kimonos were closed up, serving as pockets, as the women seldom carry a bag.

With the modernization of Japan today there is a current controversy on the subject of kimonos. Very largely the distinctive and charming kimono is being abandoned for Western-type everyday wear among the young women, though it is still continued by older women and on holidays. Not only is the kimono expensive—a good one costs about a hundred dollars—but it is heavy to wear, complicated to put on properly, and a hindrance with its tight fit and long sleeves when work is considered. Also, to clean one, it must be taken apart and remade. However, its grace and deeply rooted tradition keep it dear to the Japanese, even if it is not so generally used now as before.

Before bobbing and simple European styles were copied, their coiffures were marvelous to behold, so intricate that, once up, several days passed before the hair was touched again. The tiny wooden pillow of the Japanese, together with quantities of coconut oil used in the dressing, helped to make this possible. The hair was twisted and rolled over wire shapes, oil applied to create luster, and ornamental pins called *kanzashis*, representing brilliant flowers, were stuck in it, quite a shower of them pointing in all directions.

Today modernism has invaded many of the oriental lands. One is more apt to see *le dernier cri* from Paris or the last word from Piccadilly than the distinctively charming thing native to the land itself. This is especially true in the larger cities; one can still find tradition in the byways and small places. Today the Japanese are interested in the permanent wave, and with this inundation of Westernism, there is little doubt that complete standardization is merely a matter of time.

When the world learns to be less imitative, it will have learned a valuable lesson. There are distinctive matters of beauty that each nation has evolved. In so many ways the times are evoking greater individualism, greater nonconformity. Why not in clothes, which are so intimate and telling a part of the personality?

X I

Sleeping Through the Ages

Writers are fond of sleep, both as a subject and as an occupation, especially if they can lengthen the former and shorten the latter. Some of them, like Arnold Bennett, try to exhort us "How to Live on Twenty-four Hours a Day," in which we are assured that a cup of tea made available at a very early hour can give us a much earlier than usual start on the road to daily happiness. Some, like Mark Twain, tell us that more people die in bed than anywhere else, so it is not a safe place to spend much time in! And there are those apostles of relaxation who explain that five minutes of this practice postpones drowsiness for several hours. Finally, some dealers in mattresses and beds tell us that with their remarkably comfortable outfits three hours of sleep are as refreshing as six on an ordinary bed. Thus if the counsel of the wise is pointed toward minimizing sleep, we may at least thank the progressive furniture manufacturers who make possible such efficient sleep that its duration may be economized. Sleep, however, continues a mystery and an opportunity, and has always been a matter for us to cope with.

The temporary "other-worldliness" of our dreams has always intrigued man, and was one of the first things to awaken his sense of religion and philosophy. In sleep, our subconscious minds may proceed with their activities undisturbed by the limitations that we consciously impose in our waking hours. Today most of us have learned that the last things we think of on falling asleep are the words of command to our subconscious servants, and will be the data for our subconscious activity through the night. For this reason was the old-time warning, "Let not the sun go down upon your wrath." Auto-suggestion has become an intellectual commonplace, but we are not yet wise enough to give ourselves the benefit of that knowledge so that it may prove helpful. "If to do were as easy as to know what were good to do, chapels had been churches, and poor men's cottages princes' palaces." But that this age has come to be called efficient may be due in no small measure to our increased understanding of sleep and our improved facilities for engaging in this pleasant recreation.

How different it was in those days of long ago when man slept under the open skies, constantly alert against all the probable dangers that threatened his life. It is difficult for us who live in the twentieth century to imagine the physical discomforts that were common to those of earlier generations. The history of the bed shows us that the poorest of us today has equipment that kings of past ages did not aspire to. In primitive times it was customary to sleep upon the ground, to "camp out," as it were, in any weather. Perhaps a hole in the ground would be discovered to offer slight protection, but more often not. The house of Ulysses in Ithaca was built around a sacred olive tree whose trunk its owner had carved into a bed, so that Penelope identified her husband on his return from a very long absence by suggesting that his bed had been moved, and evoking his denial of the possibility of this.

It is curious to relate that the actual incentive for creating a raised bedstead was not to secure greater restfulness but to escape unnecessary intimacy with insects in the hot climates. Were it not for those little creatures, perhaps we might still be sleeping on the ground, and the whole basis of shelter as a necessity to civilization might have been different! When the sleeping level was raised, the Egyptians found a better circulation of air, so the custom was continued. It is not every country that has learned the luxury of a bed even today. We need not be surprised if we travel in some parts of Japan to find the people sleeping upon a padded quilt laid directly on the spotless matting on the floor, with a wooden headrest to put under the neck by way of a pillow.

The southern climes, the lands of the noon siesta, soon developed the daybed, and while we are apt to think of this as a rather modern contrivance, it is indeed one of the most ancient. Museums show fine specimens of Assyrian and Roman daybeds in bronze, ivory, and other costly materials. The wealthy Greeks liked to recline on couches, and developed beds of fiber or vegetable network to serve as a spring, which was supported by a wooden frame. A mattress stuffed with dried reeds enclosed in a cover was likewise employed. Roman beds were more luxurious than the Greek. Hangings were used decoratively, but on account of the mildness of the climate, they did not become important, and the frames which were therefore exposed to view were handsomely designed.

Beds on wheels were in use for the sick in classical and medieval times, as was also a low and portable bed called *grabatum*, with mats for bedding. This is the word used in the Gospel of St. John which was translated, "take up thy bed and walk."

The bed came in for the finest treatment under the Renaissance in Italy. Its form was distinctly architectural, reaching to the ground like a chest, of which it was indeed a direct development, with a base and roof supported by four columns carved in the manner of the times.

Interest in sleeping arrangements was naturally more pronounced in northern countries because of the climate. No southern race evolved some of the expedients for warmth that we find when we leave the Mediterranean lands, such as the big, puffy feather bed. This curious object is more like a glorified pillow than a bed, and for some reason it is made so that if it covers the upper part of the person, it leaves the toes exposed, or else keeps the toes warm and neglects the chest. There is a special art in lying under it. If you lay it like the ace of diamonds rather than parallel with the bed, you can, with luck, keep most of your anatomy decently covered.

But this must have been so marvelously comfortable when it was originally devised that ingenuity seemed to reach its climax, and apparently no one in the feather-bed countries ever thought of having it full size.

Beds at the outset had primitive simplicity for the rule. Among the humble Saxon folk, beds were simply sacks filled with straw and laid upon benches or chests. This use of straw continued for centuries and the annals of a king in the early 1400's contain the quaint instruction to the royal bed-makers that "a yeoman with a dagger was to search the straw of the king's bed, that there be no untruth therein—the bed of down to be cast upon that." The covers were the skins of wild beasts. Often the choice of a hard couch was voluntary, as these robust people had a certain contempt for a soft one. In houses of the better class the bed was often built into a recess, and the sleeping house or bed house was frequently either a separate structure or a little annex to the hall. Sometimes the bed would be fitted with sliding panels to keep out the winds. Beds of this type came to England with the Norman Conquest, but the great bed of oak with its roofed top soon succeeded them. It was placed in the center of the room and heavily curtained against the drafts that entered into the poorly constructed, chill halls.

Some of these beds were of enormous size. The famous Great Bed of Ware, referred to in Shakespeare, was built of solid oak and carved with great elaboration. It measured 18 by 12 feet and was known to accommodate at least twelve persons per night; some antiquarians have put the figure at twenty-four. Rarely did anyone sleep alone; kings themselves were in the habit of having a bed companion; Louis XIII was the first who dispensed with this custom. The promiscuousness of the bed reached its height in the sixteenth century, when hospitality was ordinarily offered in the conjugal bed out of pure courtesy!

In the families of the lower middle classes and peasantry, all slept together in one bed. In hospitals as late as 1620 the dying, the dead, and the slightly ill were to be found sharing one bed. The idea of having a separate bed for each individual was so unusual that it was considered a mark of great respect to set one apart for a guest. The beds were cumbersome and uncomfortable at best, and the addition of a few more people did not, I suppose, make them any more uncomfortable than they were to begin with. In fact, in those heatless, bleak, barnlike rooms there might have been a decided compensation. Parents and children, and sometimes dogs, could unite to produce any warmth there was to be.

No night clothing whatever was customary at this time, unless the day clothes were kept on, but this was unusual. Very often artists painted high officials lying on their bed and wearing their insignia, but this was purely symbolic, for everyone went to bed naked. This custom, which already existed in the time of Charlemagne, still continued in the sixteenth and even to the seventeenth century. The body was simply draped in a sheet or shroud. The double sheet only appeared at a much later date.

These great beds were almost like a room within a room. They were completely surrounded by curtains hanging from a canopy or tester which was suspended from the ceiling. These curtains were an essential for several reasons. Since the beds were generally placed in the great hall of the castle, this was their means for securing privacy (such as was wanted). It also modified the wind and weather, and, not unimportant, helped to eliminate the dreaded poisonous night air! The tester was of solid wood, often handsomely carved and decorated, and in effect a roof, and the headboard was also solid. Those who could afford it had an attendant to hold a light over the bed all night to keep out the evil spirits that might unwelcomely intrude upon the rest of the would-be sleepers. In those days prayers were not prayers of gratitude but pleas for mercy. Men were superstitious in a world that they did not begin to understand, and the blame was always given to evil spirits who were accordingly dodged or bribed by whatever ingenuity suggested itself. Bad dreams were always the work of evil spirits: indigestion was also their doing, and saying grace before meals was due not to a feeling of gratitude but to propitiate these unknown forces.

Bedsteads were greatly esteemed during these Tudor days, as was natural for their size and dignity. There are many heirlooms of those times that have an inflated value, however, stimulated by the assertion that Queen Elizabeth slept in them; but to have done so in all the instances claimed, she would have had time for nothing but travel from one bed to another.

Four-post beds were also to be found at this time, and were undoubtedly among the imposing pieces of Elizabethan furniture. Stately and elaborate, they were conspicuous with carving, painting, gilding, and inlay. They tell vividly of the grandeur of the stately homes of England in the days when Drake was scouring the seas in pursuit of glory or booty, when Raleigh was revealing to his friends the mysteries of the pipe and the potato, and the great Shakespeare was moving the people by the magic of his words.

In Tudor times there was often need for a secret chamber or a safe hiding place, and to satisfy this want large furniture was often designed. Sliding panels might be discovered in the wainscoting, while secret entrances through the heads of beds were not infrequently fixed in some permanent position along the wall. Often there would be a false back or double paneling, so that the secret chamber might be reached from behind the bed, and recesses and doorway entrances were made to look like pieces of furniture, whereas they were only disguised outlets to stairways or hidden rooms. People did not always go to bed sober, and with whole families sharing a bed, accidents and even deaths occurred to little children who were "overlaid." Thus came into being the low trundle bed for children.

Soon after this, transportation become more general, and lighter construction and more movable forms were required. Well-to-do people used to travel with their beds in their carriages, for uncertainty as to the cleanliness and comfort of tavern beds was great, even for people who were in no sense particular. A certain king used his bed as a bank to keep his money in, knowing that it would always be near by. When Shakespeare died he wrote in his will the wish to leave his second-best bed to his wife. This was not the slight that it sounds, for the best bed was always automatically left to a widow and was regarded very reverently as an almost sacred object.

Another type of bed was known in England, and was no doubt the pride as well as the burden of the seventeenth-century housewife. This was the "mourning" bed, hung entirely in black, probably black velvet, not even the sheets in the case of a widow being permitted to be white, and the walls of the room and all the furniture were draped in equally gloomy fashion. It is difficult to conceive of anything better calculated to deprive the bereaved of any consoling reflections she might have.

Leaving England, it is interesting to examine the luxury that prevailed among the French. Two mattresses covered with silk were often used, and fine linen sheets were used to wrap around one's self, nightgowns being also here unknown.

Somewhat later, kings began to hold court from beds which were impressive and awe-inspiring. One of the most sumptuous rooms at Versailles is the bedchamber of Louis XIV, which for many years was the center of society and government in France. This is the chamber where he held his famous audiences, and it must have been a blaze of magnificence. With its tapestried chairs, embroidered hangings, mirrors and portraits, bronzed cupids, and girandoles, its gorgeousness was without rival. The curtains of the bed were so ample that the royal occupant could safely remove his wig without those present being the wiser. It must have amused him to see the great ladies of the court vying with each other in extravagance and display made possible by the vast sums which he himself supplied.

Not only did the Grand Monarch give audience in bed, but this habit produced the most elaborate ritual about the morning toilet of royalty and nobility. The king's "levee" or rising was attended by the nicest ceremony. One courtier would hand him his left stocking, another his right shoe, and others would participate in the various details of dressing him. A certain rank and honor attended each of these functions, and surely as much thought was expended upon securing these privileges as of conducting the government itself. The great ladies of this era also received their friends in the same manner, and it is accordingly easy to understand that the bed would assume outstanding importance in the history of the French arts and crafts.

The less ostentatious and more comfort-loving Louis XV did not continue this custom of holding court in bed. The heavy draperies were no longer needed, and the natural woods, no longer concealed, were given greater decorative treatment. A large variety of beds was now developed. It being an era of leisure, the daybed with two cushions became popular. Often the architecture of the room was made to include a niche into which a bed exactly fitted. Gradually the high-post bed disappeared and instead the bed *à la polonaise* was used. This had four irons rising one from each corner and meeting overhead to form a crown or hold a dome or baldachin. Four curtains fell from this, caught back at the corners. Another type known as the *duchesse* had a flat ceiling as long as the bed itself, and this was hung all around with a valance. A narrow curtain hung straight on each side of the headboard. The bed *à l'ange*, or angel bed, had head and foot boards the same height, the ceiling shorter than the bed, and curtains draping the bed.

A custom of the bed which may be mentioned in passing was this barbaric decree: in order to avoid any substitution of the child, French queens were obliged to give birth in public. This was still in force in Louis XVI's reign, and Marie Antoinette was almost stifled to death by the crowd which had thronged into her bedchamber.

In Brittany, Holland, Scotland, and many old rural districts enclosed beds may be found. These take the form of recessed cupboards with tiers of shelves for parents and children, enclosed during the day by doors, and in this connection we may recall Goldsmith's line, "a bed by night, a chest of drawers by day." In Holland they are sometimes as much as six feet from the floor, and are reached by a step or ladder. A little shelf or platform extends over the foot of the bed, and on this the children sleep. The space under the bed is often used for curing the famous Dutch cheeses, for these are combination rooms, everything in the house going on in the same place. Our modern combination rooms know the sofa bed, which serves as bed by night and sofa by day, thus having the capacity to add the equipment of an extra room in emergencies—a real necessity for small apartments.

For artistic beauty and simplicity, few beds rival the four-poster of colonial days. Unlike the elaborately carved earlier specimens of England and France, these colonial furnishings are characterized by almost puritanical severity. We have heard of the bedrooms of our ancestors when feather beds and warming pans and heavy draperies were needed to make habitable the rooms in which the very water froze in the pitchers. In colonial homes, where bedrooms were fireless, curtains and hangings were necessary protection. Curtains were so usual an accompaniment of the bed that "under the curtain" was an equivalent of "in bed." To have curtains was to embellish them with embroidery, crewelwork, and other forms of craftsmanship which gave employment to the girls and women, and it is curious to notice that these have survived as a decorative asset long after the actual need for them departed with the introduction of better heating.

But today's living conditions show a radical difference from some of those early expedients. Where once the bedroom was large, clumsy, and stuffy, the height of charming daintiness is today obtainable. There is a new note, a freshening interest, a special fascination in modern adaptations of fine bedroom furniture, in which we may trace the lines of the old craftsmen and yet have the comfort that modern life demands. Modern beds have been equipped with steel side rails, cross rails, and corner locks, the most rigid and sanitary metal bed construction. Creaks and squeaks are gone and forgotten. Warping rails are *passé*, and the extensive rough and dusty inside surface of wood rails need no longer worry the particular housewife. Colorful bedrooms are called for, and it is a trend that commends itself for permanence. Furniture is now used not simply for utility but to key the color scheme of the entire room, either to blend or to contrast harmoniously with rugs, walls, draperies, and bed coverings. These effects are obtainable not only in the traditional fine furniture woods, but in paint, and in the fascinating metal bedroom suites whose glowing color warms their coldness and commends them for decorative interest as well as conspicuous durability. Bed coverings today, too, are of all sorts of interesting materials. Printed linens, chintzes, brocades, silks, lace, velvets, and hand-woven fabrics are all well liked and, chosen to supplement the period and feeling of the room, are a significant factor in its development.

Many modern bedrooms now include phonographs or tape recorders by which occupants do their "sleep learning." It appears that in those few moments just before falling asleep, one is especially impressionable, and the thoughts, lessons, or instructions one presents to the mind at that time have a chance to sink into the inner mind with comparative ease. Sleep learning is used by actors, opera singers, lecturers, and others who have memorizing to do, as well as those who study for examinations, or need to learn another language quickly. Habits of self-confidence, will power, and other positive character and personality traits are being cultivated by this passive, easy method.

While the modern bedroom is furnished consciously to be decoratively interesting, its primary purpose is to provide a place for privacy, peace and relaxation. From it are debarred distracting objects and, be it said, disturbing thoughts. The up-to-date person will know how to sleep promptly and refreshingly. We no longer count sheep; we clear our minds deliberately of our problems, using our ingenuity to overcome insomnia as well as other problems. We think, if at all, of something as impersonal, as remote, as wonderful as a star. We absorb the peace of some beautiful and quieting idea.

Thus we may say, as did old Samuel Pepys, "and so to bed," to rest thankfully in the comforts of this modern age.

Beds are really awe-inspiring objects. One time I was inspired to write some verses about them. They are in no sense perfect, but the ideas may interest you, and if you will not be too critical, I'll add them here.

As I travel in the elevated train,
I look into the second-story windows of the tenements.
Through nearly every window is a bed, for the poor
Have room but for essentials, and a bed is life itself.

From birth to death, we start and end in bed,
And all our lives, potential powers there are nourished,
For in our solitude and in our sleep
We reverently rise to walk with God.

No matter if the bed be but a board and rail
As with the poor, or deftly carved and canopied;
If ailing little child or one of fame and power repose,
The bed is one's own place; each has his world apart.

Beds through the ages, tossed on night to night,
Torn with the anguish and problems of the day,
Beds, human beds, of love and grief, of hope and of despair,
Beds to lament in, beds to create in,
Beds to repent in, beds to await in;

Beds that welcome night and dread the dawn,
Beds that joy in health and energy and cheer,
Beds to go to with an honest, weary yawn,
Beds in which to hide a secret tear . . .

Beds I remember seeing in distant lands,
Fair couch or downy feather bed,
Or older yet, mere planks to rise above the ground,
Evading insects in a torrid clime,
Beds built like cubicles to exclude the drafts.

Count but the dreams that through the ages
Have been dreamed in beds!
The dreams of hovel and the dreams of palace
Make men equal; for in that other-world of fancy,
We are what we think we are, and no opinions irk.

And saints and angels come as freely
To the cot to give untroubled rest
As to the cluttered boudoir with its fancies.
Asleep, no man-made yokes need bind us to their wills.

The life of Bed, that unreal, most real life,
Is veritably ours; we give the magic word of sleep
To our subconscious minds, and with the poet
"Dream true" and go where'er we will.

In long-gone days, no king did have our simplest, hardest beds,
Yet men can dream while camping in a cave under the open skies.

Perhaps the dreams are sweeter from a hard bed—
It is a pleasanter place to leave for unknown realms.

As I travel in the elevated train,
I gaze into the second-story windows of the tenements.
Through nearly every window is a bed, for the poor
Have room but for essentials, and a bed is life itself.

XII

Furniture's Forefather: The Chest

IN OLDEN TIMES the chest was not only the main piece of household equipment but often the only one. In one way or another it was able, even if a little crudely from our modern viewpoint, to serve adequately the simple needs of the people. Unspoiled by luxuries and comforts in those grim, stern times, they would use a chest not only to hold valuables, but to sit on, sleep on, eat from, and ever so many other things. Today, with our combination rooms, combination furniture, and general desire for *multum in parvo*, we should be interested in a piece of furniture that could contain the essentials from which so much of modern furniture has developed.

For the traveler in those days there was no certainty as to the accommodations that would be met with—or, rather, the certainty was always that they would be bad. Taught by experience to be prepared, he would take along those basic needs that today we assume as part of the simplest roadside service, and a chest serving the purpose of a trunk would always accompany him. The things he regarded as necessities differed tellingly from the modern tourist's luggage. Something to hang on the wall to keep the wind out, and a rug for the floor, would always be included. The great tapestries brought along would be hung in a corner so as to create a sort of partition whereby the lady and her court might find privacy. In the chest, salt and pepper and other spices would also be carried from place to place, for these things were expensive and hard to get, and were needed to conceal the bad taste of things that passed for food. Having arrived at a destination and made the place "comfortable" from the contents of the chest, the edibles would be placed upon the top of it while it was used as a serving or dining table. When bedtime came (I do not know if they called it "chest time") they would throw some rugs over the chest, fashion some cushions into a mattress effect, and add some tapestries for blankets.

When chests took the place of trunks in the early Middle Ages they were built in the simplest form—mere boxes resting flat on the ground and quite without fine workmanship. In the Renaissance they assumed decorative importance, and their form had the architectural interest of an Italian palazzo, with pilasters at the corners and a cornice above.

Such chests were given as wedding presents to newly married couples and acquired an almost symbolical character, the outcome of which was a most luxurious ornamentation. The most prominent artists and craftsmen lent their skill to the decoration of a *cassone*, which was the name for the bridal chest or marriage coffer in Italy. Sometimes they were made low so that they could be used as seats, and sometimes high so that they could be used as tables. The usual form was long and narrow. In some of the larger *cassoni* the panels were enriched with paintings in the religious style of the times; and some decorated by Botticelli, Andrea del Sarto, and other great artists have come down to us. Often they were covered with leather upon which the coats of arms of both families would appear, and historic or symbolic themes were employed upon them.

In all the intervening years, the hope chest has lost none of its popularity. It is today the ingenious custom in a certain little town in Turkey for every father to plant a tree when he is blessed with a new daughter. The tree grows up with her, and when the time comes to start the significant hope chest, the tree is cut down, cut up, and sold. The money thus derived is applied to the purchase of the chest.

If a *cassone* was intended for a trousseau, it would sometimes have small divisions for fans, lace, combs, or other feminine adornments. Then, as now, cedar was popular for keeping dresses or tapestries, as its vapors protected the contents against moths. The humble cedar chest in which the modern woman keeps her treasured woolens is a copy of the early piece of period furniture, and preserves not only the woolens but the original form of its distinguished ancestor.

Houses in those days did not have built-in cupboards or closets, pride of the modern housewife. The house without its chests would have been as little to be desired as a house today without a closet. The early armoire was an important development of the chest. This was originally used for housing the armor, but as times grew more peaceful it was used for the ordinary clothes that succeeded armor, and thus came to be one of the first forms of wardrobe. Increasing wealth in clothes called for more commodious furniture, and by easy stages we arrive at the gradual addition of drawers to the chest. Imagine the ingenuity that went into the designing of the now commonplace drawer! First, instead of the top lid or cover for the chest, there was substituted a dropleaf on the front. Then slowly a lover of convenience thought of fitting a smaller chest into the large one that could be removed and would afford greater access to the contents. Then this forward stride was improved by the adding of slides to the chest within a chest, which was called a drawer because it could be drawn out. The addition of a drawer to the chest began in Cromwellian days. When the superiority of the chest with drawers over the chest without drawers manifested itself, lowboys and highboys became known. The highboy was a chest upon a chest, the higher one set back and slightly narrower than the one upon which it rested. It was made in two sections, mainly for convenience in moving, and partly, by breaking up the lines, to lighten the appearance of what would otherwise be rather ungainly and clumsy-looking. The word has nothing to do with "boy" but is derived from the French *bois*, wood.

Not only did the ancient chest contain clothing, linens, and weapons, serving as a wardrobe, but it was also a depository for valuables before the days of banks. Sometimes an iron ring was fastened to each end and it became a traveling safe, to be slung on a pole between two mules, and thus could be ever under the eye of the owner on his journeys. It is said that merchants used them as seats when they were negotiating a sale, further protecting the contents in those none too honest days. The fact that the chest was movable and could be kept under the constant watchfulness of its owner was a precious quality in warlike times. It could be taken on campaigns and crusades, and appeared under tents as well as in rooms. It was provided with a lock and key, and contained divisions and secret hiding places which added to its interest. The ornamentation of the chest was of real importance in those illiterate times, as it assisted in identification.

Some of these treasure chests or coffers were wonderful examples of craftsmanship, with complicated spring lock, grillwork, escutcheons and key, carved bands and ornaments, all of course hand-wrought in handsome fashion. Some of them had holes in the bottom by which they might be screwed to the deck of a ship. Often the locking machinery occupied the whole of the inside cover, all beautifully designed, and complicated enough to discourage any attempts at theft. The whole idea of the coffer was strength, and if it were not entirely of metal, there would be a great deal of it in its construction in the form of iron bands and similar safeguards. The chest of that type may be regarded as the forerunner of the fireproof safe of today, and adequately served the same purposes in much more artistic form.

Chests were also the ancestors of the chair, for it was obvious and natural to sit upon them. Standing as the chests did in the state rooms and halls of Italian palaces, their convenience as seats was early discovered, and in England they were requisitioned for this purpose long before chairs came into common use. From the early boxlike chest, seating furniture evolved. A low back and arms were added, making it a box bench, or *cassapanca*, while if the back was high and ornamental, it became a thronelike seat of imposing character. Since most of the chests were very wide, the combining of them with backs and arms made them also the forerunner of modern sofas. Some of the first medieval chairs were made as small chests, with backs and armpieces as temporary additions which could be removed and put into the chest itself for transportation when necessary.

The medieval chair was built high from the ground so that the master's feet would not touch the cold stone floor. Instead, he would dangle them rhythmically in mid-air or else rest them comfortably on a footstool. But in those days a footstool was called a "cricket," and it habitually occupied a place on the hearth as the most comfortable spot in the room.

The chair is of extreme antiquity even in its modern form. For centuries, however, its use was restricted to well-born and important personages, and it became common only about the sixteenth century. Until that time stools and benches, the latter generally in the form of chests, were in wide use as seating furniture, the stool accommodating but one person, while the bench was for more than one. Old Egypt had some splendid chairs, and undoubtedly Cleopatra reposed upon pieces of ebony and ivory richly carved and gilded and laid with costly fabrics. Napoleon's Empire style is quite closely modeled on this, dating as it does from his Egyptian campaign. It is said that in old Egypt the host supplied a whole chair for each single guest, but married couples had to share one between them.

It is interesting to know that a chair without arms but with a curved back was called a *cathedra* in the early days of Rome. It was the most comfortable chair known in the Roman cities and was at first used only by women, though later men decided that they might as well be comfortable too. Its employment by teachers in the schools of rhetoric gave rise to the expression *ex cathedra* applied to authoritative utterances of every kind, and its use by bishops explains our word "cathedral."

In the Middle Ages when a family possessed a chair it was generally set apart for use by the master and became in a sense a domestic throne from which he ruled his household. Early medieval chairs, we have said, were really small chests with removable backs and armpieces. Later, as we know, these became permanent parts and were usually very high at the back and deep at the sides. This was not for ornamental purposes, as at present, but rather for protection against drafts. Such a form was essential to comfort in medieval dwellings in northerly climates, whether cottages or castles, for their crude structure and ill-fitting doors and window casings did not keep out the gusts of wind that blew through every crevice. Even the draperies hung about the walls were in many instances used for protection against these wandering gusts rather than for decorative purposes. Today our methods of building do not require such treatment, but we have learned to cherish it for its early meaning and its present beauty. What was once a bare need has become a charming accompaniment of the dignified home. The deep, high-backed chair of modern times is upholstered and made comfortable in a manner unthought of by our early ancestors.

The chair as an object of utility, then, is as old as antiquity, but as an object of comfort it is rather a modern invention. In medieval times men whose chief interest was the fight felt that mere comfort was a weakness, if they thought of it at all. A king, loving display, might wear gorgeous royal robes, while he would sit upon a narrow, hard, unyielding seat. Since furniture making of any kind was thought to be unmanly, few chairs were in existence, and these were used primarily to indicate rank and position, a tradition which we trace today in the ceremony of "taking the chair."

Gradually, with the spreading ideals of the Renaissance,

the chair became more democratic, and it began to be used among the plainer ranks of people—in fact, by anyone who could afford its possession. From that time forward it has varied in form and size to accord with the requirements of fashion at different periods and in different lands.

When the heavy armor of feudal knights gave way to simpler military attire, it was no longer necessary to have strong and heavy seats with thick supporting stretchers, so that the lighter-weight, more-open types became the mode. Next, the need for space to accommodate the farthingale and the puffed trunk hose caused another change; the side panels of the chairs were opened to make a space under the arms. In Queen Anne's day the arms were shortened and set back to allow the full skirts of women to billow becomingly about them, and very high backs were devised as background and resting place for the tall periwigs and hennins, those great cone-shaped caps that extended a foot or two above the head. Time went on, and as the French cavalier donned a costly long-tailed brocade coat, a special shape was invented called a conversation chair. A dandy could sit astride the seat of one of them, facing the back, and leaning his arms upon a shelf-like surface at the top, so that the precious tail might hang unwrinkled behind him.

Of course, the purposes of chairs and tables had been served in primitive times by rocks or stones at different levels, but these were not furniture in our present sense. The pallet of skins lodged on the sunny side of a boulder may have spelled comfort once, but not at the time of which we are speaking.

Chests were therefore soon used as tables for various purposes. Some are represented with chess squares upon them, and husband and wife sitting on them and at the same time playing the game, according to the old illuminations. More usually they held the meal and other daily utilities. As a table, the chest left a great deal to be desired, for in order to open it, one was obliged to remove everything that had been placed on the lid. Moreover, to get at anything in the bottom, it was necessary to remove the entire contents of the chest. Such a table afforded no leg room either, and since one also sat upon it at the same time, or upon another at the same level, it could not be called comfortable. Thus it was that some ingenious workmen, in order to overcome these difficulties, conceived the idea of adding legs to the chest, and then backs, and then doors and drawers, until finally the chests were no longer chests but were actually tables, beds, chairs, and cupboards.

Cupboards assumed various forms according to their special needs. Important types were the court and livery cupboards of Gothic times. The court cupboard had nothing to do with court, but was derived from the French word meaning "curtailed," or "short," and was of a more adaptable size than the large cupboard. The livery cupboard was required for holding food, candles, and other stores, which by means of open spaces were kept aired, and which were doled out to the servants and given to the poor by the holder of the key. It was called "livery," from the French *livrée*, defined as the delivery of a thing that is given—a word that survives in connection with the clothes of menservants.

The buffet seems to have been a chest elevated on uprights or columns, with a shelf below supported on feet, a more developed form of the early cupboard or the hutch.

As other needs developed, such as those that accompanied the art of writing, secretaries also planned on the same basic form were evolved. Indeed, all cabinets can trace their family tree back to chests and coffers, either directly or through a branch line.

The ubiquitous bureau is also one of the sisters of the chest family. It received its name from a coarse brown cloth with which writing desks were covered, and the word grew to mean not only the cover but the desk itself; then the office symbolized by the desk; and finally the people who worked in the office. In Europe people speak of going to the bureau as we speak of going to the office. We refer to it ourselves in some connections, such as travel bureau or information bureau. In the bureau as furniture, having drawers to hold papers, other things such as clothing were also kept. The addition of a mirror made it an acceptable bedroom piece for miscellaneous accessories of dress. So we see that most of the pieces in which we store things are at least cousins of the chest, if not more closely related. For smaller possessions the jewel case has been from earliest times an essential utility.

As a part of the proper traveling equipment of royalty, according to an old record, there was a chest for the "queene's heade." This was not quite so tragic as it sounds, for it meant merely a box for her headdress, her coronet or ropes of pearls, or her little cap of gold, studded with jewels. Her jewels she would need to have, whether there were sheets for the bed or not. Chests were then actually thought more important than beds, which only protected the person of the owner, who could defend himself, while the chest could protect his defenseless possessions. One could sleep on the ground, if need be, but one must have a safe place for those treasures which even a rather primitive person would find valuable. When beds became more commonplace they were high, clumsy pieces, and a chest was often put alongside to be used as a ladder to step up to it. Such a chest frequently proved a convenient hiding place for milady's paramour when her lord's footsteps were heard along the corridor.

Of all the forms derived from the prolific chest, none perhaps was more important or more interesting than the jewel casket. This little treasure chest was usually smaller than the other types, partly because, I suppose, people had relatively little treasure, and partly because it was made so strong that it was just as heavy as a large one.

The love of jewels is an ancient one. Only to think of the old Egyptian jewelry makes one dream of clear, starlit nights, brilliant colors, exotic brown-skinned figures of unsurpassed grace, moving about on floors of colored stone, their jewels glittering and making a tinkling sound as they walked with sandaled feet through spacious halls. In the days of Cleopatra the dancing girls wore girdles of silver and gold, wide anklets, and bracelets, all enhancing the slimness of their forms. And before Egypt was India. Charms to keep away the evil ones; tokens, then as now, from the "true love"—jewelry of one symbol or another, telling one legend or another, is almost as old as femininity.

There were beads, chains, precious stones, pins and brooches, religious ornaments; sometimes the "bride price" in days when women were purchasable property; sometimes, before the day of money as we know it, jewels were the purchasing power of the family. It is no wonder that, aside from their beauty and the fascination of their changing facets and their marvelously constant mineralogical composition, their rare color and gleam and brilliance—it is no wonder that they have been chosen the elect among treasures to serve so many varied purposes, decorative and otherwise, and that the little treasure boxes in which they were kept had so much love and art lavished upon them.

Jewel caskets have played a romantic part in literature. Do we not all remember stories like that of the lovers of Portia taking their chance on gold and silver caskets only to let Bassanio of the modest taste win the very judicial lady? Do we not remember fair Marguerite enchanted with a casket of jewels which Faust has left at her doorway?

There is something beautifully chaste about the idea of ivory for a jewel box, and much wonderful art has been expended upon caskets of this material. Many of them had subjects chosen from biblical and mythological decorations, but there were also those in secular subjects. Carvings of warriors, hunters, dancers, and animals were to be had. Wood-carved caskets were similarly treated as to decorative motifs. Often the subjects were selected from popular romances of the Middle Ages. Here may be an assault upon a castle; there is a jousting scene, or somewhere else a medieval elopement. Gold and silver have always been worthy of jewel boxes, the brusque Cellini himself having many to his credit. More spiritual is the type of copper-gilt casket ornamented with scenes of the Passion and figures of saints wrought in silver, a characteristic product of the goldsmiths of Augsburg two or three hundred years ago.

The linings of the jewel boxes would generally be of velvet or silk—in any case, something of rich, beautiful texture that would enwrap yet reveal the jewel at its loveliest. If it were of a special color, the box might be chosen to harmonize with it. A color-sensitive person would not put a sapphire into a green box or a ruby into an orange one, though doubtless the owners of these boxes did not concern themselves with such subtleties.

One of the amusing features about jewel cases is the secret compartments that many of them had: little drawers that seem but panels in the decorations; little hidden recesses concealed beneath false bottoms; unsuspected shelves. These are quite intriguing, but the owner must be astute to remember what is where. Many a precious stone has been thought lost, only to be finally remembered as too well hidden in some "subterranean passage" of a jewel case, and many a good detective story cut short by its simple recovery. Another trick of ingenuity in jewel-case construction is the musical arrangement whereby the would-be burglar may hear some haunting funeral march or "Nearer, My God, to Thee" to prod his conscience. The conventional jewel box will have interior divisions so that rings can be kept separate from pins, and necklaces will not become entangled in their own lengths. Of course, there are much simpler chestlike boxes for small trinkets. As with all other furniture, size is governed by use.

When directing the furnishing of small homes today we like to select each piece as far as possible for capacity as well as surface. Thus, instead of a regular table, an excellent purpose is frequently served by a chest the top of which may be used as a table, while it also provides valuable drawer space for various domestic possessions. Pieces of this sort range from the smallest to the amplest size, and since chests were so generally employed in past times, designers have an endless source of inspiration for reproductions. From the simple pine boxes used by the early American settlers, back to the elaborately carved walnut *cassone* of Renaissance days, it is possible to find just the desired type. Thus an entrance hall gains great dignity with an oak chest; a painted peasant type in the right bedroom means gaiety, while a carved Spanish chest of red velvet and decorative iron tracery would give just the dash to a handsome room in this mode. A chest that is low and broad does much to increase the restfulness of a large room, while a tall and narrow one serves a decorative purpose by giving height and making the upper part of the wall agreeable to look upon. The hanging of a good tapestry over a low chest contributes something of the atmosphere of its original setting, while upon a high one the placing of a fine piece of pottery or decorative metal may be used for color accent and contrast.

Through the ages, then, the chest in one form or another has held its place as an essential in household furnishings, and few pieces seem to evoke an equal amount of sentiment or romantic interest. Whether we think of it as holding the rich fabrics of an early trader, the gold and jewels of some brave buccaneer, the "queene's heade," the farmer's grain, the bride's dreams, or whether we think of it developing through the centuries to satisfy the needs of today, surely there is bound up with the idea of it a vital stirring quality that makes it verily a treasure chest of fascinating memories and rare tales.

XIII

The Possessive Circle

THE EARLIEST RECORDS tell us that marriage in savage times was a matter of either force, fraud, or purchase. Women were merchandise by the sale of whom their male relatives profited, or they were captives in war, the spoil of the conqueror, or they were stolen away from the paternal home. Primitive man undoubtedly preferred a captured wife, because she had the added value of being a trophy. She attested to his prowess in warfare, to his strength and courage in carrying her off. A horseman was always a groom, and as the "groom" came on a horse to carry off the girl, the expression "bridegroom" arose and has come down through the ages. The best man and ushers were originally those who followed the fleeing couple in order to put obstacles in the way of the pursuing family. The best man was the strongest and most helpful. In all cases, however, the possession once obtained, the bride became the property of the man who married her, and the husband was her lord and master.

The wedding ring is a relic from this ancient time. It dates from the days when the wife was regarded simply as a piece of property in the possession of the man. It was the possessive circle, the symbol of subjugation and slavery, signifying the band by which she was bound to her owner.

This matter of ownership was not so grievous as it sounds, for if the woman was not owned by her husband, she was owned by her father, and her position as wife was a higher one than that of daughter. A beautiful girl has always had a market value, and the primitive father was always willing to sell her to his own best advantage, getting the best terms he could from the applicants. It was indeed the most honorable kind of union from the girl's viewpoint. In a mere love match the man was trying to get something for nothing. This was considered stealing the girl, and unions of this sort were apt to be of short duration. But if he paid six or eight cows for her, that showed confidence. That marriage was likely to last!

Marriage by capture merged into marriage by purchase: the bride was first stolen and then restitution was made later to escape the vengeance of her people. Sometimes an exchange was made, the groom offering his own sister, or horses, cattle, or land, in exchange. Or he might work it out in service of some sort to her father.

The price of the bride varied according to her charms, her health, her rank, her condition as a virgin, widow, and such things. We are told that among certain peoples a widow brought only half the sum which a maid demanded. The Navajos of New Mexico considered twelve horses an exorbitant price for a wife. The word "wedding" betrays this stage of wife purchase. The *wed* was the money, horses, or cattle which the groom gave as security.

Brides among some early peoples were not bought in just this way, but valuable presents were given to the parents. The type of gifts, however, was not left to the good will of the would-be groom. It was definitely prearranged, and often before a satisfactory sum was decided upon, much bargaining would take place. Exchange of gifts is one of the earliest customs, always known to be a way of winning favor. In some countries, after gifts are exchanged, the marriage contract is considered complete and neither party can undo it. Often it is the custom of the land not to allow the engaged couple to see each other between the time of the engagement and the marriage, the idea being that they might elope and deprive the family of the price of the bride. Much later a dowry was given by the father as an indirect way of reimbursing the groom for his outlay in purchasing the girl.

Primitive man, always eager to dramatize, to express the joys and passions of his life, soon made a ceremony and a celebration of his marriage. And as these ceremonies were developed, it was natural that an element of religion should enter into them. The linking together of two lives was undoubtedly regarded as a significant performance. It was only to be expected that help for the future union should be invoked from the invisible powers whose works were apparent to primitive people close to nature.

It is curious that, for so long, women were held in the state of bondage that we know. Even the translators of the sacred writings seemed averse to giving woman any standing, so that instead of our reading of Father, Mother, and Son, we read of Father, Son, and Holy Ghost or Holy Spirit. Through all creation we find the dual aspect of nature manifesting equally in two directions, positive and negative, masculine and feminine, action and reaction, giving and receiving, and the like. One without the other is impossible. So it must be here. The equality of the sexes for which women must have so long been struggling is only a natural and proper state of affairs.

Even the story of Eve as born of Adam's rib, which on the face of it might produce an inferiority complex on the part of Eve's daughters, may be interpreted to show quite another truth. This, like so much of the Bible, may be taken allegorically. In the anatomy of man there is no rib missing. He has the same number on one side as on the other, and the same number as woman. Why then should that particular part of man's anatomy be mentioned as the origin of woman? The anatomy of the rib, our symbolists tell us, gives us the key to its allegorical interpretation.

The ribs are an essential part of the framework of the body, without which the vital organs would not be protected, or the body be able to maintain its erect posture.

Here in one graphic word it is clearly indicated that woman was not created as a mere offshoot, a possession of man, or merely for his companionship and pleasure, or as an afterthought. This interpretation by theologians was one of the major causes which for centuries kept her in bondage. But the obvious meaning of the symbol indicates that woman is not an appendage but an integral part of the body of humanity (represented by Adam) and the framework of society, without which its vital organs would not be protected or humanity be enabled to maintain its uprightness.

Further, according to tradition, it was the fifth rib on the left side that was used to fashion Eve. This too has a wonderful symbology, for that is the rib which covers or encloses the heart, the universal symbol of the feminine or heart qualities, the love nature, intuition, sympathy, compassion, and so on.

If this seems to go a little afield of the subject, I am sure that it will nevertheless provide an interesting point for reflection.

The advancing status of woman through all the ages is of course too large a topic for consideration here, but it may be recognized as the cause of changing modes of marriage. And yet despite modern sophistication, the marriage ceremony today reflects many of the customs that had their origin in a primitive age. Giving the bride away is a relic of the day when she was sold. The best man is descended from the strong warrior who aided in the getaway of the happy couple. The honeymoon is the period when the groom was forced to hide with the bride until her people grew tired of looking for her. The flower girls carried wheat before the bride to symbolize the wish that the union would prove fruitful.

These things and many others connected with the ceremony of the wedding remain only in the memory after the event, but the ring is a tangible actuality, and as such its history is particularly interesting.

The custom of symbolizing the marriage vows with a

ring goes back to legendary times. The ancient Egyptian maidens, if they did not actually inaugurate the idea, wore them pridefully nevertheless. A circle was the hieroglyphic symbol to indicate eternity. The circular form of ring was used by them as emblematic of the ideal marriage in which love continued through eternity.

There was a time some years ago when women occasionally burst into bitter agitation over their humiliation at being made to wear wedding rings. Such a custom, they insisted, was a brutal survival of the early days when rings were used as seals and men bought their wives like any other form of merchandise. Only then it was at the betrothal ceremony that a man paid the purchase price for his bride and placed upon her finger his ring with his personal device or seal, clearly indicating his ownership. In all the intervening centuries many unusual rings have marked the romance, the reverence, the sentiment, the mysticism, or some other emotion evoked by the idea of marriage. The meaning of marriage and the ceremony itself have undergone countless changes, but the wedding ring, like the brook, goes on forever. Whether the happy couple are joined according to the ritual of some Greek or Roman god, or in modern companionate arrangement, the ring is indispensable. The only ones, to the best of my knowledge, who do not wear wedding rings are the Quakers, or Friends.

Two women were conversing together, when one observed to the other, "Why, Mrs. Smith, you do not wear a wedding ring!"

"No," she replied. "Mr. Smith and I are Friends."

"Oh, indeed, I thought you were married!" came the answer.

Confusion of this kind is spared by the convenient use of the conventional symbol. Meanwhile, the little circle of precious metal that we know today has evolved through many changes of form and design.

Appropriate mottoes have been engraved on wedding rings since the early days of Greece. A favorite motto two thousand years ago was "Faith Immortal," an excellent sentiment for any period. Gold rings which are supposed to have been wedding rings have been found in the ruins of Pompeii. They are decorated with a design showing a man and a woman holding hands. Betrothal rings are much older in many countries than wedding rings. The betrothal and the wedding ring may have been one and the same, but it was always a pledge of love and a symbol of nuptial union. Plain iron rings were first used for this purpose in ancient Rome, and were still favored even when the wearing of gold rings had become general among certain classes of Roman citizens. It was about the second century A.D. when gold rings came into use in the ceremony of betrothal.

With the Romans a signet was set in the betrothal ring. This not only identified the woman as her husband's property—which our feminists so greatly resent—but also signified that the wife was to have the right of sealing up the household goods. Occasionally a small key formed part of the ring, with a similar significance. While it is true that at one time the seal ring secured a man's valuables as carefully as an individual Yale lock today, it was for this very reason given only to persons admitted into the highest friendship and trust. Hence, when it was adopted into the marriage ceremony it indicated that thereafter the wife was a sharer in her husband's counsels and joint partner in his honor and estate. We have a testimony to this view in the words of the wedding ceremony—"With all my worldly goods I thee endow."

The rings of olden times were made to suit the means and tastes of the lovers. While some of them were large to awkwardness, others were almost threadlike. The shapes differed as much as the size and were often engraved with portraits or other symbols. Gold, silver, iron, steel, brass, leather, or rush are said to have been used in rings for weddings. Some men wedded with rings of iron, after the fashion of the early Romans. Even the ring of a church key could be used, or a leather ring cut on the spur of the moment from the bride's glove. A story is told of a clergyman who used a brass curtain ring to marry a duke. Precious stones frequently added their beauty and interest, but sometimes the simplest materials were employed, and in emergencies or hasty marriages even straw and other commonplace materials seemed to serve the purpose.

A popular type of wedding ring in the later classic era was the gemel type (French *jumelle*, a twin) in finely wrought gold. In other words, it consisted of two gold hoops, exactly alike, so that when worn together they appeared to be one ring, yet they could be separated and one hung from the other. The symbolism of the two rings indissolubly united in one is of course obvious, but the separated rings had their symbolism too, since divorce was extremely prevalent among the Roman Republicans.

This gemel type of ring was so popular that it lasted through several centuries and is still periodically revived in Paris, though not as a wedding ring. There were, naturally, variations, each period adding a new device of some sort, until at last wedding rings throughout Europe were not only extremely ornamental but exceedingly sentimental as well —almost coy in character. This was especially true of that period known as the Renaissance, before and after the Reformation, when Henry VIII was ruling in England, and Catherine de' Medici, through her sons, in France.

Many interesting and ingenious designs of wedding rings have been preserved from the Middle Ages. They were often very large and massive. Some of them were of ornate Gothic designs, with projections reproducing the towers of famous cathedrals.

Today we naturally think of the wedding ring only in connection with the third finger of the left hand. It was not, however, always worn there. During the reign of George I of England it was not unusual for the bride to wear the wedding ring on the thumb, although it had been placed on the third finger at the marriage ceremony. Possibly this custom may have been due to the fact that particularly large wedding rings were favored by fashion at that time. In India the wedding ring is worn on the thumb. But it is not placed there for life; usually only for the days actually devoted to the marriage ceremonies. Having been worn on the thumb and on every finger at various times and places, the wedding ring came to be placed upon the third finger of the left hand. This revived an old custom derived from the belief that there was a vein running directly from this finger to the heart.

During the Renaissance, men as well as women wore wedding rings, but usually on neck chains or attached to ornamental pins in their hats rather than on their fingers. Martin Luther wore a wedding ring after his marriage to Catherine von Bora, the nun whom he had converted to his own faith. It was a gold gemel ring, one hoop of which was set with a diamond, the emblem of power and fidelity, inscribed with Luther's initials, and the other set with a ruby, emblem of exalted love, and Catherine's initials. It was decorated further with scenes of biblical history. It was at this time that diamonds began to be frequently used in wedding rings, for their remarkable hardness symbolized the ideal strength of the marriage bond. A typical inscription on rings of this time was one that said, "As the diamond cannot be cut, so love cannot be overcome."

A few years ago a proposition was agitated in London that men should be forced to wear wedding rings. Public attention was called to this question by newspaper reports to the effect that a young woman testified that she had innocently encouraged the attentions of a married man because she had no means of knowing that he was married. The custom of the husband wearing a wedding ring as well as the wife has long been the rule in Germany as well as in many other European countries. It has become a more general practice here, since the custom was revived early in World War II amid the rush of pre-induction marriages.

Everything about the ring came to have special significance. "The matter of which this ring is made is gold, signifying how noble and durable our affection is," it was said, "and the form is round to imply that our love will never end." "As gold excelleth all other metals, so doth love excel all other blessings, and as gold is set off with gems, so is conjugal love set off by other virtues," was another popular analogy.

In sincere but crude attempts to provide an enduring record of their pledge of troth, the men of Queen Elizabeth's time resorted to the use of verse engraved upon the ring. This fashion of inscribing tender sentiments or so-called poesies within the ring raged furiously from the sixteenth to the nineteenth century. Many of them were flowery and foolish and would say almost anything to make a rhyme. "As gold is pure, so love is shure," was one of these. Another was "My love is fixed, I will not range; I like my choice too well to change." Occasionally a distinctly frivolous note crept in, as in the ring of a certain lady who, upon marrying her fourth husband, chose the inscription, "If I survive, I will have five." In our time, not even extreme liberality would urge much in behalf of such sentiments as "Our contract was Heaven's act," "In thee my choice, do I rejoice," "I will be yours while life endures," or "If you deny, then sure I die."

Jewish wedding rings of this period were most unusual. Even those as late as the sixteenth and seventeenth centuries differ essentially from other wedding rings in vogue at that time. Of enormous size, many of them bore upon them whole miniature buildings, worked out in every detail with elaborate filigree, arranged in bosses and rosettes and enriched with pastel enamels. One such ring preserved in a famous English collection, for example, has a curious little building with high gabled roofs and enameled tiles, pierced by windows and flaunting gay weathercocks on its apex. Upon the inner side of the gold band is inscribed the Hebrew for "good luck." Other rings represented temples. It is believed that a marriage ring of such massive proportions was worn only during the ceremony and then stored away with the family valuables, another, slighter ring doubtless being substituted for everyday wear.

The wedding ring seems to have been a matter of individual taste before the middle nineteenth century. Couples must have spent long hours thinking up marvelous designs and inscriptions worthy of their undying affection, or perhaps they didn't, for some of them were pretty bad. Clasped hands and intertwined hearts were considered particularly smart. A typical English wedding ring of the seventeenth or eighteenth century, for example, consisted of two hands in white enamel, holding between the thumbs and forefingers a rose diamond, heart-shaped, set in silver and surmounted by a jeweled coronet.

Then for a long time there was a reaction against all the ornateness of the ring, the ambiguity and poor taste of the poesy. It was felt that unless the means of expression reflected dignity or the sincerity of the giver and the importance of the occasion, the message had better be left unsaid. Thus the plain gold band came into favor and so remained until the time of our grandmothers. The wedding rings worn a generation or two ago were heavy gold bands which to our modern eyes appear very crude. Compared with these designs, the graceful modern wedding rings indicate a great advance in taste. Today there is a revision to the more ornate ring. In Paris a recent novelty showed a ring an inch wide covering the third finger from knuckle to joint, and more than reflecting the importance of the elaborate Continental wedding by its striking display of jewels. In America there is also a return to the beautiful and costly ring, but with the addition of an austere simplicity in the chasteness of platinum and diamonds that does not lose sight of the sanctity of the occasion which it marks. Today the ring no longer represents the purchase of the woman. Around it hovers an invisible aura of protection, tenderness, and choicest sentiment, symbol always of the blessings and opportunities for service that are hers.

Vast has been the progress in these opportunities, notably in the United States. Successful in practically any profession and occupation, it is at long last recognized that women's equality with men is no longer to be imitative but to be a balancing force with men in stabilizing and bringing peace to our chaotic world. Men's work has been necessarily materialistic, harsh, competitive. Woman's has been life-giving, humane, and spiritual.

There is in fact coming into being an impressive international project dedicated precisely to this, under the name of Man-Woman Equality League. It is originated by the world-famous geniuses, Dr. Walter Russell and Lao Russell, and emanates from their majestic palace, Swannanoa, on a mountaintop outside Waynesboro, Virigina. Their claim, convincingly analyzed in their brilliant, hard-hitting books, is that world peace can only be obtained through balanced man-woman power. Because of the man-rule concept, man has dominated every institution of our social order—religious, educational, governmental, industrial, artistic and scientific. Not one institution controlled by man alone has the warmth and love it would have, and for which the world longs, if men and women controlled it equally.

The terraces of this lovely estate at Waynesboro are decorated by their sculpture, and there radiates from this place a rare and wonderful quality of beauty and wisdom exemplifying the merging forces of two internationally known geniuses into a project worthy of their powers.

XIV

Coming Into Money

IMAGINE, if you can, what this life would be without our modern god, Money. And yet money is an entirely artificial, man-made invention, and if we look back into the past we shall find that there was a time several thousand years ago when no such thing was known.

In those ancient days there was no buying or selling of goods, no making of loans or investments, no need for banking, bookkeeping, and none of the mad scramble that present conditions have foisted upon us. Then each family ordinarily produced most of its food, clothing, and other simple wants. Did a man want a sword or something that he could not grow or make, he would exchange some of his own grain or cattle for it. This simple method of direct exchange was known as barter, and the trade that was going on in the ancient world before the sixth or seventh century B.C. was practically all barter trade.

But barter was not a simple thing, and it presented many complications in its execution. The first difficulty was to find two persons having just those things to dispose of that suited each other's requirements. Many people might want things, and many might possess the things wanted, but to allow of barter there must be a double coincidence that is infrequent. Thus the owner of a house today may tire of it and may know of another exactly suited to his wants. But even if the owner of this second house wishes to part with it at all, it would be remarkable if he would care enough for the first man's house to be willing to make the exchange.

A second problem that our moneyless ancestors had to cope with was how to determine what quantity of one commodity would be the equivalent of another before an exchange could be effected. If a certain quantity of beef be given for a certain quantity of corn, and the corn is exchanged for eggs, how much beef for how many eggs, and similar puzzles, were constantly coming up. What a document of complications would a list of current prices have been, with each commodity quoted in terms of every other commodity! It was a great stride in advance when any one thing, like a pound of silver, could be used as a common measure of value. By knowing how much flax might be had for a pound of silver, and also how much corn for the same quantity of silver, our ancient merchant could estimate without further trouble how much flax could be exchanged for a certain quantity of corn.

Barter was inadequate for another reason. Many kinds of goods could not be portioned out. While such things as corn or eggs or small articles were simple to handle, as is our small change, there were things of greater size or value that could not be divided. A tailor might have a coat ready to exchange, but it would far exceed in value the small quantity of milk that he wanted. Part of a coat would have no utility, and too much milk would be a liability to him, since it would turn sour.

An anecdote is recounted of a French singer making a professional tour around the world some years ago, and in the course of her travels she gave a concert in the Society Islands. In exchange for her recital she was to receive a third of the receipts. When they were counted, she found herself in possession of three pigs, twenty-three turkeys, forty-four chickens, five thousand coconuts, besides considerable quantities of bananas, lemons, and oranges. In Paris, at the time, this amount of livestock and vegetables might have brought four thousand francs, which would have been good remuneration for five songs. In the Society Islands, however, pieces of money were very scarce, and as

Madame could not consume any considerable portion of the receipts herself, it became necessary in the meantime to feed the pigs and poultry with the fruit, and when all was said and done she had little to show for her work.

Barter still goes on in some cases, even in our modern communities, but as a rule only when its inconveniences are not felt. Domestic servants often receive part of their wages in board and lodging, and farmhands are sometimes paid by the use of a certain plot of ground. It is conceivable that money is entirely without purchase power in some places. If you go to the North Pole and decide to trade with the Eskimos, money as we know it will be meaningless and useless. But take along some gumdrops and a supply of fish hooks, and you have something to conjure with.

In the earliest days, when the hunting of wild animals was the chief occupation and the chief source of subsistence, the proceeds of the chase were the property of most obvious value. While the meat of captured animals was too perishable to be saved or to be frequently exchanged as a unit of "money," this was not the case with the skins. These were valuable as clothing, and lasted for a long time. Thus furs and skins were acceptable as money in many ancient lands, and down to the time of Peter the Great, leather money circulated in Russia. In Russia and China, tea money has also been used. Made of tea and compressed into blocks, it is sometimes serrated like chocolate bars, to be broken conveniently for change. It may be used either for money or for a beverage. Cheese in China was also compressed into cakes and circulated until someone got hungry and ate it.

As civilization advanced into a milder, pastoral form, sheep and cattle became the property of greatest value and the units of measurement. A man might receive a number of cows for a sword, and could later offer some of them for another commodity.

Cattle as money had the advantage that it did not need to be carried from one owner to another; it could walk there. While it needed attention and food, it also "increased and multiplied." But oxen varied in value, some being lean and others fat. Then, too, it was not always convenient to take cows about with one on a shopping expedition. Imagine a tourist today going abroad with such a substitute for traveler's checks or a letter of credit! Imagine a Europe of today with great stalls for cattle wherever we now see a bank!

Naturally, before there could be extended commerce, some form of money had to be invented, and the people began to experiment and to seek a standard of value. Metals soon asserted their superiority. They took little space and cost nothing for fodder. In parts of Africa a copper cross weighing less than two pounds buys a wife, as one collector says, of indeterminate weight, height, and good looks. Iron was one of the first metals to be employed. Sparta, which had private reasons for wanting to keep her people at home, made money of iron, and made the coins so large and heavy that no one could afford to travel. Money in strips with a hole to string it together was a convenient kind for carrying about before the days of pockets.

At first metals were handed about in ingots and were weighed at each transaction. Old weighhouses are still seen in some of the ancient market places of Europe. Later the money was stamped to indicate its fineness and guarantee its purity. The first recorded coins were minted about 500 B.C. by Croesus, whose name became a synonym for wealth, in Lydia, a gold-producing country of Asia Minor.

The chief reason why gold and silver became the money metals was that they satisfied the craving for adornment. It is the luster and sheen of gold and silver that caused them to be highly prized in the early stages of civilization when other ways of producing these effects were not known. The glitter of the bauble is the origin of the monetary use of the precious metals, precisely as glass beads and scarlet cloth have been used for barter by explorers among the savage races. In Central Africa, purchases are made and debts paid with strings of beads or coils of brass. An ivory merchant or traveler will lay in a stock of these just as in Europe he would carry gold or banknotes. We know that one of the great impulses to the making of Venetian glassware was Marco Polo's urge to the Venetians to make beads as currency to use in Africa and the East. We know that as late as the seventeenth century Manhattan Island was bought from the Indians for a matter of beads valued at twenty-four dollars.

Custom having established the use of gold and silver, utility continued it. These metals are singularly free from liability to deterioration. They do not rust, and they retain their luster with unusual constancy. Most important of all, they have proved to be sufficiently abundant for money use, and yet not so abundant that they have ceased to be prized.

The effect of money is to give a freedom of movement in a way that payment by barter could never do. If one receives cows, they are great, clumsy things requiring much care and responsibility. If one receives money, one may do as one likes with it—eat on it, live on it, use it for clothes or travel, or save it. Other commodities do not lend themselves to thrift as does money. Instead of men being tied to flocks, herds, houses, lands, and the like, the invention of money permitted men to change the nature and locality of their possessions very readily. In the second century B.C. people began to buy land not for use but to sell again at a profit, and speculation was first known. In Cicero's Rome there was an overabundance of money gathered from the outlying country as tribute and plunder from the provinces, and the bankers and money-changers grievously exploited it. Money was so scarce in the provinces that enormous rates of interest were charged. Bankers would borrow money in Rome for 4 or 5 per cent and lend it out in the provinces for 15, 20, and sometimes even 50 per cent.

In the Roman Forum I saw the remnants of the "exchange" of this period. It was a temple to Juno Moneta, the goddess of money, with marble floors, columns, and general architecture that would create envy in any modern bank. Coins of copper that were melted into the floor during fires may also be seen there, and one must admire the remarkable sense of design that was evidenced in these early forms of numismatic art.

Merchants during the Middle Ages could not rely upon the government to maintain the standard of value. Kings often debased the coinage again and again, to secure the means of carrying on war or paying public expenses of other kinds. Every debasement, as it left the coins with less-pure metal, lowered their purchasing power and raised prices. Many innocent people suffered and everyone grew chary of making bargains. In many countries the privileges of the great feudal lords included the right to keep a mint, and this further complicated matters. The central government would try to restrict this right, but in general the currency of medieval Europe was made up of a vast variety of coins and standards even less reliable than the king's coinage. It was difficult to pass a coin at its full value outside the locality where it was minted. Even in countries like England, where feudal coinage was put down and where debasement by the Government was exceptional, counterfeits were not rare, and the clipping of coin was a common business.

These things made the money-changer a necessary figure in the commercial world. He was to be found everywhere, even in the small towns, buying and selling the various coins in circulation. But while the money-changer facilitated payments in any given place, he was not of much aid to a merchant who had a payment to make in a distant town or country. Though the merchant could buy from him the foreign money with which to make the payment, the transportation of the actual coin was not only dangerous and expensive, but also subject to legal restriction, and if possible was avoided. Instead of sending money, the merchant would usually send some ware which he could sell to advantage at the destination and make his payment with the proceeds.

About the year 1000 and during the Crusades, the monasteries were the real bankers of the country. Enriched by the expiatory gifts of pilgrims, while the nobility became impoverished by distant expeditions, the monasteries lent money to needy princes and otherwise were the financial agencies of the communities.

It was not until the thirteenth century that bills of exchange were used to any considerable extent. Then they were developed in Italy and spread from there. In Italy, also, the money-changers developed other forms of banking. As they were dealers in money, businessmen in want of capital for their operations naturally sought it of them. The money-changers might lend it from their own stock or act as brokers and secure the money from some man who had a surplus. The short step from this to the common form of modern banking was made when merchants deposited their surplus cash with the money-changer, and he had thus a large stock which he could lend as long as he kept enough in reserve to meet the demands of depositors. It soon became unnecessary for money to pass at all in large transactions. A man could get a loan from a bank simply by having a deposit ascribed to him on the books, and could assign this loan to others as he chose to pay it out.

The characteristic danger of banking, the attempt to make a great deal of credit out of a little capital, appears early in Italy, with the resulting failures. The advantages of the banking system, however—the economizing of time and money and the great facilitating of business operations—were so clear that banking kept its place and spread toward the close of the Middle Ages from Italy to other countries.

Today in every city the importance of money in the city's business is attested by the dignified banking structures that adorn it. The impressive, prisonlike structure of the Bank of England, fondly called the Old Lady of Threadneedle Street, familiar to visitors to London, is one of the most conservative and important institutions of its kind in the world. In Paris, the Bourse is a copy of the Temple of Vespasian in Rome. The Banca d'Italia in Rome, the Borsa of Genoa, the fine exchange of Brussels, and countless others are evidences of national pride and progress. Today our respective governments have the matter of money in hand. We do not have to weigh it in the presence of the buyer; we do not have to worry over its intrinsic value, although inflation does present problems. Far from being an arbitrary thing, the science of money is one of the most exact, and the Treasury Department one of the most vital parts of the Government.

With all this, the problems of exchange are not yet settled, and a certain instinct for barter has survived, at least in Italy. It is amusing to see how every small dealer expects to be begged for a price reduction. It is ridiculous to spend precious time bargaining, and equally ridiculous to spend the first asking price of wares, which is generally three times the taking price. I found the usual period of argument about twenty minutes, whether it be a matter of five lire or fifty lire. One argues even over postal cards, and the Italian vender seems to enjoy a good struggle quite as much as final victory. It is quite an amusing thing when a group of tourists emerging from a hotel are "raided" by some of these peddlers. Often each of the group will have bought the same beads or postals from different men at the most surprising range of prices. Of course, there are a few one-price places in each Italian city, and the wise purchaser goes to these and does well.

The derivations of the words relating to money are interesting. Pecuniary takes us back to *pecus*, the word for cattle. Our word "money" is derived from the temple of Juno Moneta—Juno the Adviser—where, on account of its secure location, the Roman mint was established. Coin is from the Latin *cuneus*, a die or stamp. Many coins are so called merely from their weight, as the English pound, French livre, Italian lire. Others are named from the head of the state, as sovereign and crown. Guineas were named after the country from which the gold was obtained, and *franc* is an abbreviation of the inscription Francorum Rex, king of the Franks.

Our common expression for the payment of a sum of money is "fee," which is from the Anglo-Saxon *feoh*, meaning both money and cattle, a word cognate with the German *Vieh*, which still bears only the original meaning of cattle. Being counted by the head, livestock were also called *capitale*, whence the law term "chattel," our common name "cattle," and the economic term "capital." The dollar was first made in Bohemia, in a town called Joachimsthal. This meant the valley of Joachim. Because of the purity of the coin made there, it became a standard of excellence and was called *Joachimsthaler*. This was shortened to merely *thaler*, and in the form of *daler* and "dollar" is known everywhere as an honest unit of money.

Some of us have a guilty feeling when we consider how much we desire money. We call ourselves materialists, and are very apologetic when we have to discuss it. This attitude, however, is a waste of energy. If we remember the simple utility of money, its use as a convenience and agent, we can appreciate it for what it is, without overestimating it to the point of materialism or underestimating it to the point of carelessness.

At the same time, all of us need to be reminded that material assets are not necessarily the token of the good life, and they are likewise not the last word in security. One is often called upon to forfeit material comfort for spiritual security, for integrity, for perhaps a long view or a sturdier interpretation of values.

We must admit that although the progress which man has made through the ages has been in some ways amazing, it has been in large measure an unbalanced kind of progress. We have put our attention upon material things almost exclusively, while the humanitarian side has been sadly ignored.

It is interesting to observe a growing feeling among enlightened persons today that man's technological skills are far outrunning any humanitarian impulses with which he may have been endowed. The objective, impersonal attitude of science by its very nature tends to promote this indifference to the gentler, more humane qualities that are needed to establish a peaceful world. As this indifference continues, ethical and moral values tragically go into a decline.

We may love to be comfortable, but there is more to a civilized life than comfort and convenience. These are largely physical matters and are based on a process of competition, generally instigated by fear of greater success on the part of rivals. Mankind has attributes beyond the physical, namely, emotional, mental, economic, social, and spiritual. No matter what his comforts and conveniences, these represent but one-sixth of him, and are often developed at the expense of his more valuable qualities.

Now, we all know that the whole is equal to the sum of its parts. This means that each individual affects the total peace and harmony in the world. Peace and harmony are fed by love, kindness, helpfulness, appreciation, and the friendly, warm, humane traits. This coöperativeness, this emotional warmth, must be the next step, and the sooner we take it, the better. Man must use his ingenuity to achieve greater social and spiritual progress. He must equate his idea of comfort and convenience with an expansion of the individuality which can utilize these comforts and conveniences in sensible, humane, and harmonious ways. He must spend his money for something beyond the passing pleasures and follies of the moment.

Life has glorious possibilities. They are not, by a long sight, all concerned with comforts and conveniences. History tells us that the ancient Greeks had houses without drains, beds without sheets or springs; they told time without watches, crossed rivers without bridges, sailed seas without a compass; they studied geography without maps, poetry without books, politics without newspapers. Handicapped, yes; but they knew themselves as people—living, thinking, vitalized people, not small cogs in big machines. They had little comfort, but possessed a rich and glorious civilization of mind and spirit.

Let us keep our sense of proportion about comforts and conveniences. While they do not have to be the main foundation of our life, we can accept them with gratitude for the virtues they do possess.

Money, when all is said and done, is merely the symbol of compensation for work performed, and we all have heard that compensation is a basic law of nature—that we reap what we sow. The word "finance" itself shows this. It comes from the same root as "final," and is the ending or completing of an undertaking by giving the compensation or returning its equivalent.

Some of us may remember the alchemist who guaranteed a secret that would make gold artificially. The secret was to beat the yolks of three eggs for an hour without once thinking of the word "hippopotamus." The alchemist of today need not try to transmute a common metal into gold, but may transmute himself to be worthy of success. And by success we do not mean merely money, but those things that have a real value of which money is the symbol.

If someone were to give you a formula by which success of this kind was obtainable, would you follow it? It is interesting to know in this connection that Henry Ford, whom most of us concede to have been successful, devised such a formula, which I hand you for what you can do with it. This formula is not a trick but rather a mode of life. It is easy enough to understand, but since it presupposes real earnestness in the desire to achieve, it is not easy to carry out. It is said to be effective in the attainment of any legitimate desire, and consists in substance of the following parts:

1. Know exactly what you want. A hazy idea results in scattered purpose and misdirected energy. The more exactly you know, the more exactly you can plan for its attain-

ment. If you want financial success, sense what the public wants. There is no harm in large sums of money if they are kept at work opening up lines of opportunity and service. The only harmful money is that which lies idle or is used to block progress.

2. Want your success hard enough. Not eagerly today and indifferently tomorrow, but with real force and power in the wanting. There is no success without application. This means concentration of mind, labor of hands and brain.

3. Expect it confidently. Waste no energy or thought in doubting. Do and want what is right for you to do and want. Dishonest men, by obeying the other laws of success, have won a place, but it is becoming harder and harder to do that. They may have been dishonest in dealing with men, but they cannot be dishonest with materials. If you have not faith in the virtue and importance of the thing, your powers are scattered and you are putting in with one hand while taking out with the other. You must have courage that keeps you going, no matter what happens.

4. Be persistently determined to attain. Be willing to sacrifice lesser interests in behalf of the greater.

5. Pay the price by untiring perseverance. Give the equivalent of what you hope to get in work, in service, in diligence, in ideas. There must be knowledge. This is within every man's reach; there is no favoritism here. You must know all there is to know in your particular field, and keep alert for more knowledge. The least difference in knowledge between you and another man may spell his success and your failure. Guessing does not go. Trusting to luck is folly. "Going it blind" is taking a chance that may prove disastrous. You must know, and your knowledge must be the up-to-the-minute kind.

This advice is not much easier, perhaps, than the alchemist's idea with the hippopotamus. There is not the tenacity in most of us that will prevent our faltering at one or more of the parts. But without my assuming the rôle of the much maligned mid-Victorian, I think that these ways to success, tried and proved, help us feel that we are not creatures of fate; the cure for difficult conditions is largely in our own hands. To the extent that we are ready to use it are we assured of success.

X V

The Long Road to Cleanliness

AFTER TWO WEEKS of seafaring and the daily salt-water baths that one gets on board ship, one of my first ambitions on reaching my hotel at Naples was to have a regular bath. I reached Naples late one evening, and the nearest bath was down a long corridor from the room assigned to me. But I was entirely pleased when I saw it—a large room with a huge bathtub of fine marble. The room was beautifully frescoed and brightly lighted. A chair stood near the tub with a big sheet draped over it. Just as my wish was beginning to be fulfilled, all the lights went out, and there I was in pitch-black night except for the faint glimmer of the sheet. They did not light up again, and I had to manage as well as possible and grope back to my room by the flicker of candles from transoms on the way. This was my first adventure on Italian soil, and while I was glad it was of such a minor character, the incident made me wonder if it might not be better to have a plain porcelain tub with a reliable light, rather than a marble tub and no light at all.

But one soon finds that marble has always existed in Italy in remarkable profusion, and is prodigally used for the most commonplace purposes, while large-scale lighting, as we have seen elsewhere, is a relatively modern thing.

Before long, as we range through Italy, we are confronted with relics of ancient baths that combined utility with exquisite art, and we are bound to ask ourselves the meaning of the great public works represented by such structures as the Baths of Diocletian, and of Caracalla, and others in Rome and elsewhere throughout the erstwhile

Roman Empire. As we compare our good American baths with what we find abroad, it is interesting to speculate on the development of this world-wide institution.

In the history of the human race, bathing practice of one sort or another reaches back to the earliest times, and in its long career has passed through a number of curious phases. Thousands of years ago, among the Egyptians, it was carried on as a religious rite. Man, it was thought, should present himself pure in body as well as soul when he engaged in the service of his god. This thought, in modified forms, was carried down on and off until the days of Charles I of England, when a knight similarly symbolized his purity before his initiation into the Order of the Bath.

The Greeks are credited with having introduced the first bathtubs. These, however, were not occupied by the bather, but served only as a receptacle for the water which an attendant applied to him. The tubs stood about thirty inches high and resembled large punch bowls of polished marble. Among these people it was considered effeminate to use warm water, so the bathing luxury that later developed in Rome was unfamiliar here.

But the Romans had a different idea. Their baths were called *thermae* because of the warm water. When bathtubs were introduced, the wealthy had them in their homes, and facilities suggestive of the modern Turkish bath were developed. Sometimes several rooms in a house would be devoted to bathing purposes. The custom was not to be monopolized by the rich, however, and public bathing places were organized. Caius Maecenas was one of the first to construct public baths at his own expense.

These enormous *thermae* contained theaters, libraries, and all the other accessories imaginable to make them real social centers for the people. The principal ones were those of Agrippa, erected in 21 B.C.; of Nero, in A.D. 65; of Titus in 81, of Domitian in 95, of Commodus in 185, of Caracalla in 217, and of Diocletian in 302.

At first a much greater delicacy was observed in matters

of promiscuous bathing than later. But soon the public baths became places of meeting and amusement where not only did men bathe together in numbers, but men and women stripped and shared the same bath. This custom continued until forbidden by the Emperor Hadrian in the second century, when the sexes were segregated.

As wealth increased, the public baths were furnished and decorated with a grandeur and magnificence that can hardly be realized today. Every luxury and almost every dissipation went on in them; but, paradoxically, a woman who took a bath at night was considered immoral.

The baths were closed when any misfortune befell the Empire, and the Emperor Caligula made it a capital offense to indulge in the luxury of bathing when any religious holiday was being observed. Roman bathing establishments came to be huge things, and their magnitude may be inferred from the fact that the Baths of Diocletian could accommodate 3,200 persons at one time. Visitors to Rome will remember the famous collection of sculpture in the Museum of the Thermae which today occupies a small wing of these baths. The *thermae* of Caracalla accommodated 1,600 persons at a time. It was this building which served as a prototype for the great Pennsylvania Railroad Station in New York.

That the Romans demanded extraordinary magnificence in the edifices they frequented was typified in these Baths of Caracalla. Their area extended over a surface of more than a mile. The walls were splendid with marbles from Egypt and Numidia, harmoniously disposed. The ceiling vaultings were covered with rich mosaics. The water taps were of silver. Statues were everywhere. The baths contained vast halls, each of which had its own special use and name, according to its purpose. There was the undressing room, and a room where perfumes and oils were applied. There were successive chambers to pass through where one must take cold, tepid, and hot baths, then to the "sweating rooms," and returning by way of the hot, tepid, and cold baths once more, each process being attended by its ceremony. All of the chambers would be wonderfully frescoed in gay colors, and ornamental plaster was to be found in the decoration along with the marbles, mosaics, and precious metals and jewels.

At first devised for health and exercise, the baths became places of voluptuous pleasures. As a general thing, baths were taken on a "when needed" basis, but at the height of Roman luxury, men of leisure sometimes took six or eight baths a day. The regular time for bathing was after exercise and before the principal meal, but authorities tell us that these debauched creatures bathed after eating as well as before, to hasten digestion and promote appetite for fresh delicacies. Their craving for luxury was revolting to a normal person. They bathed only in hot water perfumed with rarest fragrances. They wore a ring on each finger, and so much care did they bestow on things like this that they bathed in light-weight ones in summer and heavier ones in winter.

The custom of taking cold baths became general after a doctor had cured Augustus of a severe malady by this means, but it quickly fell into disuse when the same doctor, by reason of a wrong application of cold water, caused Marcellus to lose his life. Pliny the naturalist took cold baths in summer, either being read to or dictating the while so as to lose no time from his letters; and so, too, did other scholars. At a much later date, Rostand, the French dramatist, found that only by remaining in his bath could he secure himself from interruptions, so he did much of his writing there.

The Roman baths allotted large spaces for gymnastic exercises and games. For those who did not care for these there were shaded porticos for walks, as well as galleries, libraries, and conversation rooms in which philosophers would hold forth and authors would recite their works. The baths came to be community and cultural centers, a tradition preserved in the health spas of today, such as the

An artist's restoration of the Baths of Caracalla, which covered an area of about a mile and accommodated 1,600 persons at a time.

Kurhaus of Wiesbaden, where music and other entertainments are beautifully provided. The technical skill of the Romans, as displayed in rendering their walls and the sides of reservoirs impervious to moisture, in conveying and heating water, and in constructing flues for the conveyance of hot air through the walls, was of the highest order.

While all the baths in Rome proper are far too ruined to be intelligible, an excellent example may be studied in Pompeii, though of course on a scale far less grand, as befitted the smaller community. Here, in Pompeii, I saw to fine advantage the sense of fitness that is so admirable in the decorative efforts of the first century. The *frigidarium*—a central pool for the cold-water bath—is located in a round room whose walls are a rich Pompeian red and blue, with black and gold to supplement—a warm, fine effect—and the most delicate little painted figures perform athletic stunts just as one would expect in a place devoted to exercise. This rooms leads into the lukewarm room, the *tepidarium*. The "lockers" for the reception of clothing were little cubicles all around the room at eye level, and the dividing partitions were sculptured persons acting as columns to hold up the ceiling of the shelf, each one different, each one beautiful. Then came the *calidarium*, or hot room, which is extremely interesting for its central heating system. Floors, walls, and ceiling were hollow, and the heat entered at the floor and heated the whole room evenly. Is there anything new under the sun?

Much earlier than this, in fact more than 4,000 years ago in the lost Cretan city of Cnossus, a mighty palace stood. In this palace the first "modern" bathrooms so far known to history have been uncovered. When the king of Crete felt like taking a bath he had only to step down from his throne and enter a small bathroom connected with the throne room. The queen had a bathroom attached to her hall in the domestic quarters of the palace. The water supply and the drainage system of this Labyrinth were remarkably complete. King Minos was undoubtedly proud of them in the twentieth century before Christ. He need not have been ashamed of them if he had been present when they were dug up in the twentieth century after Christ. Good as they were, however, they were limited to this little island, for there was no international communication, so that subsequent developments in other lands were independent growths, meeting the needs of the people as they were recognized.

The cutting down of the aqueducts by the Huns and the gradual decrease of population contributed to the decay of the Roman bath. Bathing, however, was kept up in full vigor in the East. Hot baths and especially hot-air and vapor baths were adapted by the Moslems, and the Arabs brought them into Spain. The Turks at a later time carried them high up the Danube, and the Moslems introduced them into Persia and India.

The modern Turkish bath is a direct descendant of the ancient Roman bath, as Turkey became the seat of government of the Empire when Rome fell. Remains of Roman bathing resorts, often very gorgeous, can be traced all along the Mediterranean coasts, in Africa, in Algeria, the Pyrenees, the Carpathians, the Alps, and Auvergne. The beautiful little Cluny Museum in Paris contains the remains of a sumptuous warm-bath establishment erected in the first century.

In contrast with the warm bath of this period, the cold river bath was the hardy German method, and when Roman dissipation halted on the arrival of the Teutons, baths ceased to have the importance they had known.

As the Dark Ages progressed, a new attitude prevailed over Europe—one that was, if possible, even worse than any that had preceded it. Both clergy and laity acquired the superstition that cleanliness was contrary to the idea of holiness, and that filth and squalor were manifestations of piety. To the Christian fathers this life was only a preparation for the life to come. The body was mortal and to be despised; the soul was immortal and to be cherished. What need to wash this miserable body which holds the soul away from heaven? So bathing became uncommon. Homes and peoples became filthy and sick, and one scourge followed another.

Until the fifteenth century, at least, community health and sanitation were beyond common experience. One might any day walk along the streets and see someone dump slops out a second-story window, even from noble or royal dwellings. This habit made walking impracticable in the streets, especially in the side streets, lanes, and alleys. Fine ladies and gentlemen rode up to the door of any place they wished to enter. As for the peasants, no one considered them, and they did not apparently consider themselves.

Hygiene and sanitation as we know them today were unheard of. The flies which swarmed everywhere were considered only a nuisance. If some wise man ahead of his time had traced an outbreak of sickness in castle or town to a polluted well, he would have been considered the worst kind of radical or infidel. A man might get the plague from having his refuse drain into his drinking water, and wonder why God should so afflict him.

To make matters worse, an ingenious labor-saving device hid the decaying refuse on the floor: rushes were spread on the stone pavements of the castle halls, and these concealed the bones, the bits of food, and all the other rubbish that was thrown on the floors. Occasionally the old matted rushes were covered with new ones laid on top of them.

To counteract the evil smells which arose everywhere, even in kings' palaces, lords and ladies had their scent boxes, which they sniffed delicately when an evil smell crossed their noses. Even the men in velvet and the women in silks and satins did not make any pretense at bathing, and for these there were strong perfumes to disguise the odors of uncleanliness. It was partly to fill the scent boxes with rare perfumes that navigators like Columbus risked their lives to discover new routes to the Indies, the home of spices and perfumes.

Refuse was allowed to accumulate wherever it might fall,

and Mark Twain only a relatively few years ago said of his travels in outlying districts of Germany that the wealth of a man was indicated by the size of his manure pile, for the more horses and cattle, the more manure on display.

Here the question arises, Where were physicians all this time? Did they have no interest in sanitation? This remained unanswered until the Crusades brought East and West in contact, and the people of England, France, and Germany became acquainted with the methods of the East and gradually revived under its wholesome influence.

For many ages scientific sanitation was undreamed of. At first the offices of the priest and the physician were combined in the same person. One who was believed to have power over demons to expel disease-producing spirits stood in the place of the doctor. As the medicine men of savages and the priests of early civilized peoples wanted to increase their influence, they were ever stimulated to acquire knowledge of natural actions and the properties of things, and thus secured useful information. Later, the writings of Moses contain allusions to the practice of medicine among the Jews, and the purpose was largely aimed at promoting cleanliness and preventing contagion. One can find very specific instructions as to sanitary behavior at that time in biblical records.

But the real art of medicine was first cultivated in Egypt. The medical school at Alexandria, founded 300 B.C., was remarkable, as here for the first time a standard of professional knowledge was set up. The greatest of the Alexandrian anatomists, Herophilus, is said to have conducted his vivisections upon the bodies of condemned criminals in order to learn the facts about the human body. Others opposing this method developed the science of drugs. This spirit of inquiry lasted but a century. Meanwhile it had been introduced into Greece and was taken up by Hippocrates, who is regarded as the founder of the modern science. It reached Rome about 200 B.C.

It was not, however, through the Romans that the modern world received its medical science, but from the Arabs, whose conquests had brought them into contact with Greece. Between the fourth and seventh centuries A.D. Arabian culture was in a brilliant golden age. In medicine they made great advances over Greece. They studied physiology and hygiene, and their drugs and medicines were much like ours today. Their surgeons understood the use of anesthetics and performed some of the most difficult operations. And so it was that, through the agency of the Crusades, Arabian culture was brought to the West and disseminated through Europe, where up to this time the practice of medicine had been all but nonexistent.

During the Dark Ages it had been mixed with astrology and superstition. It was believed that the position of the moon, the movement of the stars, and even the hour of the day affected the body. It was also believed that touching the relics of the saints brought cures, and for this and other reasons physicians were held in low repute. There were three classes of medical men: the regular physician, the surgeon, and the barber who was permitted to treat bruises and minor ailments.

A decidedly inhibiting note in the progress of medicine was the fact that the clergy so largely interfered. The Church expected cures to be effected by religious rites performed by the clergy, and more or less effectively prohibited the practice of medicine. Thus the knowledge that physicians had inherited became mixed with absurd errors, and little effort was made to verify it in the Middle Ages. The theory that disease arose from natural causes that might be discovered and removed was denounced. People felt that whatever happened was the work of God, and it was irreverent for man to be inquisitive about His affairs. Before the devastations of the Black Death in 1348, sanitary measures were, as we have seen, almost unknown. Then men began to clean the streets, filter the water, and take other measures against disease.

In the matter of bathing, even this began to be taken more seriously, and we find that the steam bath commended itself. This was especially noticeable among races that wore fur or wool next to the skin, as in the northern countries; and those who could afford it had the necessary equipment. England did not use baths much until Shakespeare's time, but even then the public baths were much more popular, as they were places of good fellowship, much like taverns or clubhouses. Games were played, beer drunk, and a good time was had by all. It is said that Queen Elizabeth bathed once a month whether she required it or not.

Soap, though known to the ancients, was then merely a hair-coloring device. And even at that it was a rare and expensive luxury, used only by the rich. In the thirteenth century olive oil was found useful in making soap for cleansing purposes. France took the lead in soap manufacture, and in the fourteenth century England followed. In 1622 King James I granted a monopoly to a soapmaker. This fortunate man was allowed to make three thousand tons of soap a year, but had to pay the king the equivalent of $100,000 for the privilege. As soap was considered a luxury, a luxury tax of a penny a pound was put on it in England in 1711.

The social aspects of the bath were still the dominating ones at this time, when the mighty and élite of England joined in making the town of Bath the most brilliant social center of the kingdom. It was then that Beau Nash, Oxford man, gambler, and exemplar of fashion, drifted into Bath, saw its possibilities, aroused the city to its opportunity, and, on the crest of the wave which he succeeded in producing, swept into power as master of ceremonies, social arbiter, and uncrowned ruler whose word was law to the throngs that went to Bath. As an example of ingenuity, it will be of interest to see how this was accomplished.

Nash was born in Wales in 1674. His father, a small glass manufacturer, wanted to make a gentleman of his son, and it must be said that he succeeded far beyond expectation. Richard was sent to Oxford, but though he exhibited undoubted signs of genius, they were not of the scholastic variety. His industry took the form of winning ladies' hearts, conducting pageants, and trying his luck at the gaming table. In these pursuits he considered a uniform an advantage, so he became a soldier. His expenses, however, were large, his income small, so he sold his commission and became a law student. His ease of manner, ready tongue, and his careful dressing always attracted attention. So when the master of ceremonies at Bath was killed in a duel, it surprised no one that Nash was offered the succession.

At this time Bath, which had been famous for centuries for its healing waters, was a very crowded and squalid place. Everything was on a small, mean scale, and nothing was done to attract visitors by developing the town. Nash took hold and in various ways succeeded in launching it as a real resort, taking advantage of its natural possibilities, installing music, entertainment, fine buildings, and an "atmosphere."

Impeccable in dress and manner, he flattered the women, an avocation which cost him nothing, and provided revenue-producing gambling rooms for the men. As civic reformer and promoter of idle pleasure, he manifested a charming graciousness and rare dignity. Having succeeded in these efforts, he now undertook to correct the provincial tone of the place. His code of manners was a model of its kind and left its imprint for all time on English society. To read it even now is to be impressed with the ingenuity, wit, and personality of its creator.

All this exerted an influence far and wide. Those who had been drawn to Bath only in pursuit of health now found fashions flourishing—fashions in art and drama as well as those more sartorial. Sheridan's masterpieces were not regarded as absolute triumphs until they had received the favorable verdict of the critics in the little Orchard Street Theater, which still stands, though used now for other purposes. The spacious houses once occupied by Clive, Nelson,

Pitt, Gainsborough, Garrick, Reynolds, Jane Austen, Fanny Burney, Mrs. Siddons, are still there for us to see, and to hear echoes of a golden age. Richard Steele, Gay, Pope, Congreve, Lady Mary Wortley Montagu, Fielding—their shadows still people the place. Sheridan, Goldsmith, Johnson, and Boswell resorted to the waters. Lord Chesterfield wrote some of his letters from the little town where worldly politeness might be studied; Walpole some of his where gossip so abounded. De Quincey was a schoolboy in Bath; Southey spent his childhood there; Coleridge, Burke, James Wolfe, conqueror of Canada, and many others added to its fame.

It is not strange that in Bath, as late as this glamorous period, conditions were little better than in unmoral Rome, and much wanton debauchery was described in the annals of the time. The discovery that these places were means of spreading diseases gave them a setback and revived interest in the home bath. More than this, the general use of linen for underclothing curiously brought a change in bathing habits, for by wearing clean linen next to their bodies people seemed to feel comfortable with less bathing!

The course of cleanliness followed a similar road across the Channel. France in the Middle Ages remembered the custom of frequent ablutions that the Romans had introduced into Gaul, but the barbarian invasions, razing their *thermae* to the ground, seemed such an insuperable obstacle that bathing simply ceased as a national habit for many centuries. Here again it took the Crusades to revive sufficient interest for a resumption of personal cleanliness. In the large medieval French towns baths were re-established and became numerous, prosperous, and immoral. In the thirteenth century a decree was issued forbidding that the same establishment be used by both sexes. The proscription was doubtless not observed for long, as in the fourteenth century hot baths were again ill-famed. In the fifteenth century the Church preached against their impurity. In the sixteenth, feasts and banquets took place therein. By the end of the seventeenth they had fallen into such disrepute that honorable people abstained from entering, and preferred to sacrifice even the small amount of bathing that was at best possible.

Only in recent times were two bathrooms installed in the palace Quai d'Orsay, when the President of Czechoslovakia was visiting in Paris. Heretofore there had been no bathrooms at all in this palace, although it had at various times been the headquarters of Edward VII, Czar Nicholas, Albert of Belgium, and others. But while a bathroom was lacking, there were tapestries, sculptures, and decorations that are really treasures of art.

As for that, it is a matter of record that absolutely no sanitary conveniences of any kind existed even in the palace of the Louvre or at the court of Versailles in the sumptuous, luxury-loving days of Louis XIV. People were supposed to take their precautions before coming indoors, and it is known that they used the rows of statuary as retreats when necessary. While soap may have been a luxury, perfume most certainly was not!

The primitive habits of certain sections of Europe are even today an endless source of surprise. On my first ride along the roadsides just beyond the busy parts of Naples, I was charmed with the wayside shrines, of which there were many examples, and I took great pleasure in nudging my neighbor as we approached them in a swiftly moving hack. Imagine how I felt when I suddenly discovered that some of these were not wayside shrines but wayside water closets, quite as open and blatant as you please, after you once identify them. As they are freely patronized by men of all classes, one is not long in knowing what they are. In sophisticated Paris this problem is met even more conspicuously. As Anita Loos tells us, when Coty came to Paris, he took one sniff and realized that something had to be done! On the very choicest locations of the main streets horrid retreats sprawl themselves malodorously. They have partly open tops, so that a tall person's foreheard is visible just over the upper edge. One rainy day I saw a man and a woman approach one of them. The man went in, while the woman stood outside obligingly holding an umbrella over him. There is much more that could be told.

The question of sanitation depends largely on water supply. Water supply outside cities has always been a great problem. Most villages in European countries are very old. When many of the houses still inhabited were built, some two or three hundred years ago, a bathtub piped with water had never been heard of. Public fountains or public wells supplied water for the community, and old European towns are conspicuous for the wells and fountains on which the famous sculptors of the Old World have lavished the best of their love and skill. German fountains tell the interested observer the legends of romance and history, of fairy tale and fact. Italian wells speak of the grandeur of ancient days. The people carried water from these into their houses in pitchers or jars balanced nicely on their heads.

This state of things still exists in many rural communities. There are still places in outlying sections of Europe where modern pipes and plumbing have not yet found their way; where a small quantity of water has to be warmed in one room and carried up to another that is, likely as not, cold and damp and unheated. This condition naturally discourages frequent ablutions, and the people in such communities feel quite noble in adhering strictly to "Saturday night"—a recurrence of the symbolic, religious attitude of being clean for Sunday and for church.

To take a bath under these circumstances is more than an event. It is a ceremony. In most hotels in France a *grand bain*, a great bath or bath in a tub, is, in all but the most modern, charged for extra and is almost a function of state. The *valet de chambre* usually acts as grand chamberlain and conducts the bather through long chilly corridors to where the bath steams in an enormous tub. The English, fond of bathing though they have become, do not seem to take very kindly to bathrooms as we know them here. Many of them still cling to the bowl and pitcher, even though a bathroom is right next door to their own room.

In the slum districts of some of our large American cities,

in spite of the existence of bathtubs in certain tenements, the tubs are used to grow vegetables or flowers in, as these are of more interest and importance to some immigrants than water.

A bath was a rare experience in the early days of America's struggle for existence. To begin with, there was no water in the house, and preparation of the bath was a task that took a long time. Water had to be carried in by the bucketful from the brook or, if one was lucky, from the back-yard pump, heated in the tin wash boiler, then poured into the wooden washtub. The smaller children could sit in the tub, but poor father had to stand up, dip a sponge in the water, then squeeze it out on his chest. That was the origin of the shower.

Bathrooms that bore any resemblance to those of today were first looked upon as a wild eccentricity of earls and noblemen, and then as a sinful extravagance, bringing a maze of messy, leaky pipes into the house to loosen the plaster of ceilings, flood the floors, and freeze up and burst in winter. No wonder they were not popular! The miracle rather is that without any medical knowledge or propaganda to support bathing as a desirable habit, intuition nevertheless came to the fore to promote and simplify it. Eighty years ago the American bathtub was a boxed-in wooden structure with a tin or copper lining. Pipes and other fixtures were closed in behind or beneath so that dust accumulated without restriction. The same equipment was common to both the plain and the expensive house, and if distinction was required, it always took the form of more carving or paneling in the woodwork, and the emphasis was laid on ornament rather than a consideration of sanitary conditions.

As common as bathing now is in America, far outstripping every other country in this respect, it is a matter of surprise to learn that bathing was, in addition to the other supersitions, originally considered dangerous to health. This was believed not only by the masses but by people of education and culture, who also strongly opposed it. Preachers denounced it from the pulpit as ungodly, and it has been alleged that the town of Boston passed an ordinance forbidding the use of the bathtub except on the advice of a physician, with the penalty of a fine or imprisonment. Though some say this was a hoax, it contained enough truth to reflect conditions.

Today we care for neither the dissipation of the Romans nor the other extreme of medieval ignorance or Renaissance indifference. Our bathrooms are among our most cherished possessions, but we have learned to use them rationally and wholesomely.

That the bathroom is the shrine of the modern American home has been humorously satirized on the stage. The scene shows a Park Avenue apartment with the rental sign indicating prices somewhat as follows: 6 rooms and 6 baths, $20,000; 5 rooms and 10 baths, $40,000; no rooms and 24 baths, $60,000.

To be sure, some puritanical souls would look upon all this as devastating luxury—which of course it is. Doubtless they believed that Americans had more character when they washed at the old farmyard pump. But anyone who has washed at a farmyard pump for long is generally willing to let character go and take the bathroom, though not the $60,000 variety.

Surely most of us would choose our present hot-water supply and sanitary conveniences as a balance for all the other luxuries that the centuries have evolved. The development has been a slow one, and in many cases mother wit has seemed to sleep inert. There is not much of constructive value to build on in the past history of cleanliness. At best we can merely profit by its mistakes.

But now there is the awakening. It is a cause for gratitude not only that today we have the scientific knowledge that enables us to theorize on the subject, but that recent ingenuity has been able to provide the actual facilities whereby we can profit by the knowledge in practical ways.

Nowadays much popularity is given to the shower bath,

the basic theory being that one bathes in constantly clean flowing water. Quite recently there has come to public notice a modification of this which has allegedly sensational results. It is alleged to be a real fountain-of-youth experience, and all it takes is time. To do it properly one should remain under a warm shower, about 110 degrees, for about eight hours, playing it in turn over every part of the body for about ten or fifteen minutes at a time, and repeating this continuously. One need not stand, but can sit or lie under it. Here one may profit from a combination of three valuable factors—moisture, warmth, and gentle massage—with passivity. A series of these baths has been found to strengthen and energize, to relieve and even cure many sorts of aches and pains, to reduce weight, to rejuvenate. Experimenters have declared themselves to feel years younger. My source for this information is a book by the well-known nature healer Dr. Benedict Lust, *The Fountain of Youth.* For those who cannot give eight hours at a time, units of two hours can be tried, though of course results would take longer. According to the description, it provides inner as well as outer cleanliness.

Whoever discovered that cleanliness is next to godliness was perhaps as ingenious as the more technical inventor!

XVI

Overcoming Distance

IN OLDEN TIMES, travel as we know it today was nonexistent. To travel for "pleasure" was a thought unborn, for there were dangers and hazards at every step. A man traveling alone was a suspected person and might suffer horrible cruelties, for there was little law to protect him. One knew nothing, except for a few monstrous legends, of the rest of the world in which one lived. In fact, we today know more of the world as it was about 500 B.C. than anyone living at that time could have found out.

Early trade, always a powerful motive for travel, was largely on land, and the dangers to life and property led merchants to band themselves together for mutual aid and protection. Such a company was termed "caravan," from a Persian word meaning trader or dealer. The hardy camel, of course, was the ship of the desert, willing to carry a load of seven hundred pounds or more over a distance of twenty miles a day, with only a pound of coarse food and a pint of water; one could hardly seek a more gracious slave.

The earliest means of carrying goods from one point to another was on the backs of human beings. Many devices were made for economy of service and strength in carrying. Bands over the shoulder and over the forehead added something of comfort and security. The whole process of the development of transportation was based on the impulse to be relieved of this bodily service. First were devised simple rough sleds on which heavy loads were dragged. Later, instead of pulling these primitive carts themselves, the men trained animals to do it. The Plains Indians had a method of attaching two poles, one at each side of an Indian pony, the poles extending backward and dragging on the ground. Upon them was built a little platform on which goods were placed and carried from camp to camp.

One day, feeling in a receptive mood, an observer saw that the rounded stone without projections moved easily, and discovered that the heavier things could be rolled more readily on a fallen log which was also rounded. Such logs were then put under the sledge. Later, circular slabs of wood were cut from the trunks of trees to serve as solid wheels, and then the wheels were attached to form part of the carriage itself. The original invention of the wheel was undoubtedly as revolutionary a thing in its influence as any of our modern wonders. In various forms, materials, and locations, regardless of transportation, it enters into practically every mechanism in use today. It was soon found that the wooden wheel could be made lighter without losing strength. Crescent-shaped holes were made in it, and ultimately the spoked wheel developed.

Since the dog was the first domesticated animal, he was early harnessed to a rude sled or drag by means of which he pulled articles from place to place. The Eskimos have used dogs and the sled very extensively. The camel, the elephant, the llama, the horse, and the ass also were used as beasts of burden after their domestication. Enormous packs were placed upon them, and still are in certain lands.

But ancient land transportation was insignificant. With such inadequate aids, all distances seemed long. Since trade was by poor roads, across deserts, or through mountain passes, and since it was dependent on human porterage or beasts of burden, it consisted largely of articles of small bulk to which high value was attached.

Greek roads led chiefly to sacred places such as Delphi, and their economic importance was small. The Roman roads were established for military rather than economic reasons, but were vastly more extensive. "All roads lead to Rome" was not a metaphor but a fact during Roman supremacy.

Centering at Rome, they extended not only through Italy but, under the Empire, to Constantinople, Asia Minor, along parts of North Africa, west to Spain, over the Alps to Gaul, and through Britain. The Via Appia was the Roman Lincoln Highway. Begun by Appius Claudius in 312 B.C., it may still be used.

These roads were amazingly built in straight lines, disregarding rights of property and obstacles of nature alike. Elevations were cut down, depressions filled, streams crossed by means of stone-arch bridges. The roads themselves were constructed to be proof against all seasons and weather, and many of them are still in a remarkable state of preservation. They did a great deal for ancient transportation, but medieval neglect has served to deprive them of their full usefulness. The Roman who traveled by land had a better time of it than our own forefathers of the thirteen colonies. His inns were not equal to theirs, but his vehicles and cattle were fully as good, and his roads were the best that have ever been constructed.

Nevertheless, the cultured Roman of the first century cared little for travel. For fame or fortune no distance was too great, no hardship too severe, but sight-seeing as we understand it had no appeal. He was perhaps indifferent to natural beauty that had to be sought with such difficulty, and because of the lack of facilities for ready communication, he felt that to be away from Rome was to sever every desired tie. For rich young men whose talents promised a brilliant future, the training of the schools was often supplemented by travel. But the world was small, and few places were visited outside of Greece, Rhodes, and Asia Minor.

By land the Roman used simple vehicles drawn by horses or mules, while for short distances sedan chairs or litters were employed. Senators, patricians, and other upper-class Romans traveled with a great retinue and piles of baggage. Antony's journeys were like a traveling circus, with his huge coach, the carriage drawn by lions, and the ostentation of gold. This luxury was far exceeded under the Empire. Nero always had a suite of a thousand carriages. His mules wore silver shoes, his muleteers scarlet liveries, his outriders and runners were gorgeously clad. Poppaea, his enslaver, had her horses harnessed with gold and had five hundred asses with her so as to bathe in their milk every day.

Slavery and the lack of decent inns account for these elaborate traveling arrangements. The rich man, with hundreds of slaves, could carry his palace with him everywhere, and service had become habitual to him. Local innkeepers could not cater for such luxurious tastes, and it was not possible to buy en route such necessities as the wear of travel made it desirable to replace.

The first tourist of antiquity, probably, was the Emperor Hadrian, in the second century. He visited in every Roman province at a time when the Empire was at its greatest extent. In all these places he took a peculiarly keen interest in architecture, and brought back with him actual plans of striking buildings as well as the usual fund of tourist recollection. His biographer reports that Hadrian traveled with companies of artists and architects, recruited and marshaled after the manner of a legion of soldiers, and few of the provinces were left without some evidence of his architectural taste and munificence. At Tivoli, a fascinating suburb of Rome, he aimed to make an architectural record of his travels. Just as today we might create a scrapbook of snapshots or other pictures, Hadrian built real buildings, duplicating famous Greek temples, Egyptian structures, and other significant architectural edifices. There were also the countless accessories of a royal residence—schools, theaters, baths, and all the rest—scores of the most pretentious buildings. Just as the excavations in Pompeii help us to reconstruct in the mind's eye the life of a Roman community in the first century, so do we acquire from the ruins of Tivoli a similar bird's-eye view of the next hundred years.

Quite as ingenious as ancient land travel was transportation by water in early times. Boats were made very early indeed by people of the lake and river side. At first they were no more than trees and floating wood, used to aid the swimmer. Then came the hollowing out of the log, then the combining of logs into rafts, and then, with the development of tools, the boat took form.

One of the most ingenious modes of transportation I ever heard of, and one of the earliest, is recorded by Herodotus after a visit to Babylon about the fifth century B.C. The boats, says he, are circular and made of skins. Their willow frames are cut in Armenia, and on these, serving for hulls, a covering of skins is stretched outside; thus the boats are made, quite round like a shield, without either stem or stern. They are then entirely filled with straw and their cargo is put on board, after which they are suffered to float down the stream. Their chief freight is wine, stored in casks made of the wood of the palm tree. They are managed by two men who stand upright in them, each plying an oar.

Then comes the interesting part. Each vessel has a live ass on board, vessels of large size having more than one. When they reach Babylon the cargo is landed and offered for sale, after which the men break up their boats and sell the straw and the frames. They load their asses with the skins and thus set off on their way back to Armenia! The current is too strong to allow a boat to return upstream, for which reason they have collapsible boats. On their return to Armenia they build fresh boats for the next voyage.

The art of navigation was taught by the Phoenicians to the ancient world. Their ships were long the accepted models of construction. While the beginning of the ancient sea voyages is lost in the obscurity of the past, we know that they were highly developed by 1500 B.C., when Sidon was the leading Phoenician city. It is said that Necho, the king who ruled Egypt about 600 B.C., sent out Phoenician sailors to attempt the circumnavigation of Africa; and the same king took up the work of cutting the first canal across the isthmus of Suez. The canal was allowed to fill up with sand, but was reopened later, and its course may be distinctly traced, it is said, along the route of the modern canal completed by de Lesseps in 1869.

The ancients, we have pointed out, soon began to realize the necessity for commerce. With regard to such things as food staples, a lack of commerce had two serious results: A good year brought a surplus for which there was no market outside of the village where it grew, and which could not be used for lack of skill. A bad harvest, on the other hand, meant serious suffering, because there was no opportunity to buy food supplies outside and bring them in. Famines were frequent in one place while near-by surpluses existed, unknown to the sufferers, and famine was often followed by pestilence. Such conditions promoted transportation in all lands.

Among the Greeks and the Romans the type of boat used was called a galley. In their heyday these were wonderfully swift vessels, but they were also terrible prisons for the rowers (generally slaves), who sat in crowded tiers tugging at the oars. Biremes had two tiers of oars, triremes three, and quinquiremes five. There are said to have been galleys with as many as sixteen tiers of oarsmen, one above another. This human engine went rhythmically on and on to the beat of a mallet on the deck, while a soldier with a heavy whip acted as oiler, according to Arthur Warner, applying the lash to the bare backs of the rowers whenever he thought the machinery needed lubricating.

In the Mediterranean a vessel with sails might lie becalmed for days together, while a vessel with oars would easily traverse the smooth waters, especially with coasts and islands everywhere at hand to give shelter in case of storm. Oars, therefore, became the characteristic instruments of navigation, and their arrangement was the chief problem in shipbuilding. As long as the Mediterranean nations dominated Western Europe, vessels of this type were built even in the North, though in that section there was too much wave for oars and wind enough for sails. Thus sailing vessels became inevitable.

But while vessels were no longer unusual, there were no transportation companies, no lines of boats or vehicles running between certain points or offering a regular schedule at a fixed price. If the traveler lacked the means to permit him to buy or charter a vessel for his exclusive use, he had to wait at the port until he might find a boat going in the desired direction, and then make such terms for his passage as he could.

Then, too, the boats were small and very uncomfortable in rough weather. The lack of a compass caused them to follow the coast as much as possible, and this generally increased the distance. Moreover, coast dwellers were overfond of practicing the rights of wreckage. Fishermen often set up false beacons to lure ships to their doom. They sold their victims as slaves and appropriated all the property.

No one except for matters of great urgency risked going to sea in winter. During the bad season ships came home or harbored abroad, for under the circumstances navigation was difficult at best, and travel by water was, when possible, avoided. Voyages were made by preference on clear, starry nights. The steersman piloted by the stars, to whom the passengers made worship before starting; but alas! in a storm, just when they were needed most, the stars would be invisible.

Not only was knowledge limited, but the unknown regions were by superstitious imaginings peopled with monsters and destructive forces. The real perils of the uncharted seas were augmented by organized bands of pirates, led by men of high rank who regarded the business as no disgrace. Then, even as late as the thirteenth century, there were the strand laws, according to which a ship became the property of the owner of the coast upon which it might be wrecked. Lighthouses were few, and false signals were made to increase the profits of this business. It would seem as if early sailors plundered all they could, and traded when they had to.

The merchant traveling by land also suffered even more

from bad men than from bad roads. Government was so weak that robbery was common. People were so ignorant of everything outside the narrow sphere of local interests that they suspected every stranger, and too often there was reason for it. A quaint law was passed in England in the Middle Ages relative to the muddy and rutted highways of the day. According to the law, bushes and trees were felled for a hundred feet on either side of a road to prevent the gentle inhabitants from rushing out and attacking travelers! Security in travel was directly helped by the Church, and the Peace of God gave to religious pilgrims and to traders a protection which they could not otherwise enjoy. An Anglo-Saxon king secured by treaty from Charlemagne the rights for religious pilgrims to have safe and free passage through his dominions, but those who were traveling for gain were to pay the established tolls. It is not surprising that many traders assumed the guise of religious pilgrims in order to escape these payments.

The rise of towns and the establishment of fairs made a certain amount of land and water transportation necessary, but it was slow, insecure, and costly, and subject to burdensome and arbitrary tolls. Only the rich could afford to pay for the transportation of wares over any distance, and they would spend their money only for valuable luxuries that were worth the trouble.

In the Middle Ages the neglect of roads and the return to more primitive methods put rather a stop to any ambitious traveling. Its gradual revival came about through the stimulus given by the Crusades in arousing interest in commerce with the East, the consequent growth of the Italian cities, and, later, the formation of the Hanseatic trade league in the towns of northern Europe.

It is curious to note that, lacking compasses, the mariners of the North steered by heavenly bodies and the flights of birds, and they sailed boldly on the open seas. Crows were taken on voyages, and from time to time they were liberated. If the ship was far from land the birds, in the manner of

Noah's dove, would return to the deck. If the ship was near land, however, the birds would fly for it in a straight line, and the captain would then steer his ship in the same course.

While the compass was known to the Chinese, it was of little use in the form that they had devised: it was unfixed, and consisted of a magnetized needle supported by a splinter of wood floating in a dish of water. Only when it was mounted on a pivot and placed over a dial in a box did it become valuable. Such improvements were made in the fourteenth and fifteenth centuries, the credit for the compass's introduction into navigation being given to the seamen of Amalfi.

The introduction of the compass made transoceanic voyages possible. The discovery of America made them profitable, and helped to shift the center of trade, and hence of travel, from the Mediterranean to the Atlantic, from Italian cities to Portugal, Spain, and the Netherlands. The destruction of Eastern caravan routes, because of Turkish and Mongolian invasions, gave special significance to the discovery of an ocean route to the Indies made by Portuguese navigators in the fifteenth century.

Sailing ships in one form or another were the queens of the seas for many centuries, and our own early seaports could tell us many a vivid story of the vessels that set out from these shores. Typical of these, in Massachusetts, were the rise and fall of Salem's shipping, which constitute one of the most romantic and important chapters in American history. Chivalry, daring, and actual heroism, mingled often with tragedy, have lent color to the immortal exploits of Salem sea captains who steered their eventful courses through the waters of the seven seas. Not only did these men, most of whom were mere boys in their teens, pilot their ships safely over countless jagged coral reefs and through many a narrow inlet, but they were entrusted with the problems of purchasing new cargoes to take the place of those discharged. Without charts other than those made by themselves, and with the aid only of crude instruments, they carried the name and fame of Salem to foreign lands in every quarter of the globe until New England became famous as a shipping center throughout the world.

Imagine, if you can, a more picturesque sight than that which Derby Street presented during the era of Salem's shipping prosperity. Ships from Africa, China, Sumatra, Zanzibar, India, Batavia, Arabia, and other distant lands docked at the old Derby Wharf laden with the wares of the Orient. This street was always thronged with sailors from all corners of the earth, typical sea rovers of that day. Monkeys, parrots, and various other animals and birds were brought here, to say nothing of curiosities of every description. Old Salem with its rare colonial doorways and arching elms is still reminiscent of these days of prosperity. We can still see the deck roofs with their railings around the edges which served as lookout stations for the shipowners to look out to sea to catch the first glimpse of a home-coming vessel.

But the efficiency of these clipper ships was not yet approached on land. After his Russian disaster, Napoleon traveled from near Vilna to Paris in thirteen days. The distance was about 1,400 miles, and the time averaged under five miles an hour. But Napoleon was traveling with every possible advantage and did the trip in about half the time that a less-favored traveler could have made. This was no better than was possible to the first-century traveler between Rome and Gaul.

But now came a tremendous change. Railways arrived, and thereupon this same journey could be accomplished in less than forty-eight hours. The chief European distances were reduced to about a tenth of the time they had been. Since then railways have assumed all sorts of duties. Down the mines and up the mountains, across the plains and through the wilderness, they puff with equal nonchalance. And as for tunnels through rocks and mountains to shorten the distances, why, just try the journey between Florence and Venice, in which you go through enough marvelous tunnel engineering to suffice you and all your grandchildren.

The Harry T. Peters Collection, Museum of the City of New York

The Clipper Ship "Nightingale" Getting Under Weigh Off the Battery, New York, a lithograph published by N. Currier in 1854 after a painting by J. E. Butterworth. The clipper, built in 1851 at Portsmouth, New Hampshire, was named after Jenny Lind, who at the time was making her great American tour.

The story of railways and their importance in unifying our own great country is too well known to tell of here.

The wanderer through Europe today constantly enjoys experiences in travel that, while modern and convenient, are, each in its own way, unique and reminiscent of the ancient times. Small wagons with a horse and a goat, or an ox and a mule hitched together; the glorious white oxen of Umbria moving the hay; the great Belgian dogs guiding the little milk carts; the fascinating two-wheeled cart of Palermo, with its bright paintings and decorations, climbing the hills to the tune of jingly bells on the mule's collar; the picturesque tallyho with four horses that took us up from Interlaken to the Wetterhorn glacier and gave us the beautiful ride through the majestic Alpine country. And of course the Venetian gondolas. They need no description here, but one cannot think of travel in Europe without a word of tribute to them. They are all black now, as it appears that when they were variously colored, political factions used them as party colors and caused disturbances of one sort and another, hence a municipal regulation for uniformity. They are very distinguished-looking, with gold and silver trimmings, long, graceful lines, and the perfect comfort they afford. The brilliant atmosphere and color all around prevent them from seeming in any way somber. Each house has steps leading into the canal, and a post in front, like a barber's pole, to hitch the gondola to. As you sit in the low gondola and glide slowly along with not a sound but the swish of the water, you cannot believe that a New York or London belongs in the same scheme of things.

While the large cities of Europe have most of America's worn-out taxicabs available to take you around, London has its buses, and these figure conspicuously in the appearance of the city. About a hundred bus routes, reaching in as many different directions through London and its immediate suburbs, are the rather overwhelming mode of best seeing the wonders of that small universe. They are covered with advertisements that one does not resent, for the simple reason that one feels so intelligent to be able to understand them after months of travel in non-English communities; they are positively flattering.

One can sympathize with the tourist who was forever traveling in the wrong direction. In Rome he asked one of his party, "If I go straight from here, how far is it to the Vatican?"

"Well," was the reply, "if you keep straight on the way you are going, it is nearly twenty-five thousand miles, but if you turn around and walk the other way, it is about a mile and a half."

The present has been known as the Motor Age. Not only for travel but for so many industrial uses as well do we rely on the motor, that we can hardly picture the world of today without motors whirring in and through our multitudinous mechanisms. Let us consider a few items concerning this aid to speed and convenience.

In order to make even the simplest kind of a motor, more than a hundred different materials are required. Many persons work with hand and brain to build electric motors, and other thousands far away, in a score of different countries, labor collecting necessary ingredients. Some are dressed in strange garb; they speak a strange language; they dig in mines; they harvest in fields; they pick and gather and labor in many ways to provide the bare materials. Then these materials must be carried across the world on land and sea, on elephants' and camels' backs, and on peons' shoulders, before they arrive at the electric works in our own country. For the copper wires that carry the current inside the motor, the ore is mined and smelted in Montana, refined in New Jersey, and drawn into wire in various towns in New York, New Jersey, and Connecticut. Some of them are covered with silk made by yellow men who labor in the Orient. Others are covered with a compound made of rubber that brown men tap from the trees of Ceylon, mixed with zinc oxide from Missouri, and other chemicals from France, Sicily, or Louisiana. Still other wires defy the heat with a covering of asbestos mined in Quebec. The wires are wound into coils and covered with cotton tape. The cotton was picked by black men in Texas. . . . And so it goes. This is only a part of what concerns itself with the copper wires. The same kind of a story could be told for the iron and steel and other factors. The international quality of our finished manufactures is an inspiring story of coöperation and organization.

In all travel, whether foreign or local, traffic regulation has always been a problem. In ancient Rome the roads were so narrow that coaches were only permitted the person of wealth for a grand occasion. Sedan chairs were used for short distances. In Amsterdam today traffic is forbidden in the evening on the main street, Kalverstrasse, where the whole town comes out for its walk and overflows beyond the sidewalk. Naples is a large city and a crowded one, but there seems to be no traffic regulation at all. The cabs go neither right nor left but wherever there is a space, and how it happens that all of Naples isn't killed off in a month remains a mystery. The horses have gay collars, usually with bells, and these, plus the honking of horns on the ubiquitous automobiles, create an incessant racket. So we have the streams of carriages and carts hurrying madly along, the drivers yelling and cracking their whips, tassels and flowers twined together on the harness, mixed teams of horses and mules or goats and horses. The confusion suggests a circus or menagerie or anything but a city engaged in the serious pursuit of commerce or industry. Italian policemen, so far as I could see, are exclusively decorative, and even in such emergencies as would give them an opportunity to be of service they stand by as one of the onlookers and simply add color to the scene. There, if ever, one thinks of the man who said a pedestrian used to be one who walks, but today he is one who runs or jumps, usually in front of a motor vehicle. In New York it has come to the point where the father says to the little son, "If you will be a good boy, I'll take you downtown to see the accidents."

We have heard a great deal about cars running down

pedestrians, but daily there are pedestrians who are knocking down automobiles. There are several major types of such persons.

One of them is the sleepwalking pedestrian. His thoughts are at home or at work, or anywhere but where he is, so that he walks along in a daze and wakes up to find himself in the middle of the street with horns honking, drivers cursing, cops glaring, and maybe an ambulance near by.

Or your technique may be to go halfway across, get scared, then take a few crablike steps backward. This is like the man running to get out of the rain and falling into the creek.

Other effects are the kangaroo, the grasshopper, and the frog, willing to dash into, over, around, or in front of any moving vehicle. And we mustn't forget the goat. Many pedestrians in our big cities seem to prefer being the goat.

Again, you may fancy yourself the essence of grace darting and dashing across the street in and out of traffic from behind parked cars or otherwise. You may arrive safely on the other side with all the airy lightness of a ballet dancer. But think of the cost to the poor motorist of your journey through space!

To preserve your bones in one piece while you save a second or two, Johnny Driver makes a short stop. He saves your life. You get over safely. "I made it," you brag, no thanks to you. But what happens to him and his passengers?

Sudden short stops are the most frequent cause of traffic accidents. They lurch passengers off the seats or knock them sideways through glass windows or into steering wheels, partitions, and other rigid parts of the car. All kinds of broken arms and legs, crushed skulls, cut wrists, displaced hips, and other major injuries occur to the driver and his passengers, laying them up perhaps for weeks, causing loss of earnings, often leaving permanent deformities, and sometimes resulting in loss of life. Often a series of cars behind must stop suddenly to avoid collision, thus causing a whole chain reaction of costly calamities.

The average pedestrian seldom realizes the damage he

does, even when he gets to the other side unharmed himself. But in addition, there is bound to come a time when he will try this once too often and, in his sprint through space, will fall short of his destination and end up with a long and painful reminder of his own selfishness and stupidity.

Traffic congestion being what it is, making your Grand Tour across the street is often as much an adventure as a flight to Europe, although it does not cost as much or take as long. You should dress for it. Use shoes that won't fly off or turn over, hats and umbrellas that won't cut off visibility; and know the language of road courtesy.

If you zig when the driver expects you to zag, you will give him not only such dangers as already mentioned, but a light touch of apoplexy as well. This isn't so healthy and has a tendency to detract from his poise in making a clear-cut job of running over you. He may just bounce you around like a football and hurl you over the hood. This hurts you more and longer, and does not give as much sympathy or cash to your widow and orphans as the clear-cut job.

As a pedestrian, you may think the driver is a pretty careless fellow not to see someone of your importance when you are right under his nose. However, as you realize if you are driving, that is not all he has to do. He may not have room to move out of your way. There may be, and usually are, cars in the lanes at either side of him, and others breathing down his neck just behind, as well as some high-steppers in front who may decide to rival your antics by turning right or left unexpectedly. Once you get caught in a hole like this, you may have to take it lying down.

When you cross a street, don't be the one to cause a sudden stop. Remember it is not only your own life and health that are involved. Allow the driver enough time to see what you are doing.

Don't practice jaywalking. You may become perfect at it, and will have your accident that much sooner.

Don't cross in the middle of the street. You double the hazards when you play both ends against the middle.

If you are going to be a sleepwalker, *à la* Lady Macbeth, be sure you have not killed anybody first. Cross with a clear conscience. Just crossing over should offer you enough emotional distraction. Do your heavy thinking at home.

Leave your troubles on the sidewalk. Never cross a street with them. You may be satisfied that they will be on the other side waiting for you, every last one of them, so you don't have to add others on the way over.

If parked cars bar your way, emerge from between them in the most cautious manner. Let them be your armor until you have peered out in all directions. Be especially careful when two lanes of traffic are proceeding in the same direction. You may get ahead of the first or nearest car, but it is the one just beyond that may get you. The driver cannot know that anyone is dashing across, and it is not his responsibility to expect you in a forbidden area when he has the right of way.

Let your actions be decisive so that the driver knows what you are going to do. If *you* don't know which way you are going and you start teetering forward then backward, what kind of a genius does the driver have to be to dodge you? Indecision on your part is strictly anti-safety. Whatever you intend to do, do it consistently. Save the jitterbug for the dance floor.

If a policeman is there giving instructions, see that you watch, and act accordingly. He is the symbol of orderly procedure.

Above all, remember those traffic lights. They are there not to be decorative but to tell you in unmistakeable terms exactly what to do or not to do and precisely when to do it or not to do it. Traffic lights are not there to delay or annoy you. They are your greatest safeguard and protection. Cooperate with them, and they will truly guide your path. Disregard them, and it will be too late for ingenuity.

And what is wrong with starting out a few minutes earlier so that you can stand on the sidewalk and wait for your traffic signal like a dignified human being who is master of his fate, and not a dasher hither and yon, controlled by impetuosity and playing havoc with himself and others?

Aside from overcoming distance by hop, skip, and jump, there is, of course, travel in its broader sense.

Travel has wider purposes today than trade alone. Exploration, archeology, scientific investigation, new ways of safety or of profit, are always tempting adventurous spirits. Byrd in Antarctica might find a deposit of something or other that may revolutionize an industry. The clean-cut flight of Lindbergh across the ocean completely enthralled the world, and certainly did an incalculable service in stimulating progress in aviation. Beebe under tropic seas, Roy Chapman Andrews finding prehistory in the Gobi Desert—all these are travelers in the finest sense. Exploration is in a very different style from some of the methods we have recorded. Your modern explorer is an expert and his results go to experts, whereas the early seafarer was more a rough-and-ready adventurer who took a long chance hoping for gain, and did not mind much if he lost. It has been pointed out that while Columbus spent only about $2,000 to discover America, Byrd needed more than $1,000,000 to enter the Antarctic. On his North Pole trip he spent nearly $200,000; yet all he did was to make a seventeen-hour trip over the polar sea by air. Martin Johnson spent $250,000 on his last expedition to Central Africa. Few modern explorers are able to take a large scientific staff into the field under a cost of $100,000. The fabulous sums now spent for adventuring into space, for atomic-powered submarines, put even these large sums into a casual category.

With telegraph, cable, radio, Weather Bureau, and countless similar services, it is a simple thing today to find out the conditions in any particular place and to sally forth by motor, train, ship, or plane. We may be secure en route and may give ourselves unreservedly (as long as the travelers' checks hold out) to the lure of unfamiliar places. Today traveling is a fine art, and is practiced by dilettante and professional with equal ease. Fast trains and steamships have annihilated distance. Regular routes at short intervals take one from place to place in safety and security, and with every reasonable comfort and luxury, while airplanes of all kinds roar overhead and bring the world closer. Indeed, so delightful are the means of travel that to many it has become the chief goal of leisure. Tourist companies are organized to care for all the details and mechanics of travel, leaving one free to attend only to the objects of the journey, whether business or pleasure. The traveler today returns not wearied or worried but renewed and refreshed by his experiences.

The influence of transportation has been reflected in the appearance of our cities. First the cowpaths through the wilderness, or the little rock landing for a small raft. Then the great harbors and enormous railroad stations; now the landing fields and the rooftop garages. The largest railroad station in Europe has long been that in Leipzig—not that this city is the first in size, but because of the great fairs held there which draw untold numbers of people. It is an industrial and cultural center of real importance, and its railroad station indicates this by its great size.

As I was seated in my train, having left the station, an American girl in my compartment said to me, "I want to get off at Leipzig; can you tell me when we get there?"

I replied, "Why, that was Leipzig; you've passed it."

"Is it possible?" she said. "Why, the sign said *Ausgang*." Unfortunately, that was merely the German word for "exit"! So the young lady had to ride to the next station.

Consider the building of roads, of tunnels, of bridges, of canals, the charting of the seas, the surveying of the lands, the adventuring into outer spaces—what things have been done to facilitate our travel! See in the mind's eye the Indian's canoe, darting down some sylvan falls, and then the raft, the rowboat, the sailboats and clipper ships, the steamboats, and now the great floating hotels of the sea and the air. Survey the long road between the ancient sedan, the two-wheeled chariot, toboggans, sleds, and sledges of the snows; the howdah of the elephant, the desert camel; the medieval coach plodding through its dreadful roads; the covered wagon bearing our pioneers to the Coast; the bicycle and the motorcar, the subways and our "rabid" transit, railroads and airplanes—what a story of ingenuity, of adventure and romance, of overcoming obstacles and giving service!

The purposes of travel have become almost as numerous as the methods. Satellites are barely conversation pieces now, and what they tell us of weather and what not can grow continuously more significant. Jets to the moon intrigue the interest vastly, but one wonders if there are not enough projects, more needed, closer at hand. Atomic submarines are now objects of national rivalry, as, regrettably, are so many of our technological advances; whereas, because of modern facilities for going to places, travel could become more and more an agent of interracial friendship and understanding. While the United Nations labors to put these ideals more firmly before us, the world is slow to follow these paths of wisdom. Surely with the proper good will, we find by visiting in other lands the innate, normal heartiness of people everywhere. We recognize our prejudices for the prejudgments that they are. We realize the human strivings of folk in other lands, and find that we may become neighbors in spirit as well as in actuality.

XVII

Man Becomes Literate

Man is an evolving creature. Given nothing but his great desire for happiness, or for self-expression (much the same thing), the creative urge develops within him, and he finds ways to adjust himself to life, to add powers and facilities that he did not start with.

It is difficult for us to imagine a time when man, who can talk so incessantly now, had no language. But generations after his advent he gave expression to his feelings in rude grunts or cries, as did his brothers among the lower animals. A few gestures helped him out.

But while man was and is always an animal, that is only a portion of his nature. Latent within him, very deeply buried perhaps but somehow there, is the feeling that he is "made in the image" of the Great Spirit. And so he builds and builds from those inner reservoirs of power.

We may look back to those prehistoric centuries and say, How simple life was then! But was it? Would you like to find yourself on a strange planet, with nothing that you needed or wanted, with no precedent for getting or doing, with—well, why picture it all? Man had a tremendous job ahead of him. And it is wonderful to see how, not even able to realize the purpose or the goal, little by little he worked at each problem as it occurred, left it when the solution seemed reasonably satisfactory, to give his attention to something more urgent; came back to it as cause or opportunity presented; improved it, neglected it, carried it on; and thus began, oh, so slowly, to sense what it was all about.

So, then, man first came, unable to talk; but the power was in him to develop speech as he realized the need of it.

Man was a long time growing merely articulate, before such a thing as graphic representation of an idea occurred. In those earliest days men passed their beliefs or their knowledge down from generation to generation by word of mouth, committed to memory. It was perhaps the head of the tribe, more often the poet or chief priest, whose memory was the repository for the tribal literature, and who declaimed it before the campfire.

Such priests or poets existed early in the far-off days. They developed among their hearers the sense of something greater than themselves—the tribe. They saw the life that even then extended into the past. The desire to be remembered was born. Efforts to keep popular legends unchanged were more successful than could be imagined in times when written records are common. But variations did creep in, and it was decided that memory needed some tangible help. In order to mark the place of a decisive battle or some other great event, trees were sometimes planted or piles of stones gathered. Later we find that the bark of these trees was scratched with symbolic pictures, and carved or painted monuments were made of stone or wood. Egyptian obelisks and Alaskan totem poles are historical documents of this kind.

But as experiences increased, there was much more to remember, and gradually a rude system of writing grew up. First, pictures of objects were used. A circle stood for the sun; a crescent for the moon; the idea of a tree was shown by the picture of a tree. As man grew in knowledge, certain fixed signs began to indicate certain sounds, and a phonetic system of writing developed. The hieroglyphics, or holy writings, of the Egyptians represent one of the earliest of man's attempts to reduce human speech to graphic form. The hieroglyphics are a scheme of picture writing in which familiar objects, such as animals, birds, and the like, represented certain thoughts. Taking the idea as it would work out in English, if the primitive person had the conception "treason," it would sound like "tree-son" to him, and if he wanted to write it he would put down the symbol for tree, and a boy, which would represent son. In other words, it was rather a process of charades. Such pictures, carved on clay, brick, or monuments, made eternal records of what the Egyptians thought and felt five thousand years ago. Modern Chinese writing is still so pictorial that China has never been able to make the slightest use of the conventional telegraph because the language does not lend itself to the use of code. Only since the invention by which the original telegram can be photographed and sent thus has China been given a device by which its whole trade has been revolutionized.

But practice in picture making inevitably led to simpler methods. Soon it was seen that really only part of a picture would express an idea just as well. If the idea of sorrow was to be denoted, a weeping eye was as intelligible as an elaborate presentation of a man in tragic pose. The alphabet was a gradual growth. Certain sounds appearing in various words like (if we were using English) "bee," "because," "behave," "bead," and so on, would all be written with a single symbol for the "be" and the addition of the symbol for the second part. Thus the symbol for "bee" would grow into a letter with always the same use. After a long process of evolution, it would be found that a limited number of symbols, usually about twenty-four to thirty-six, could represent all the sounds in the language, and that these symbols could be combined independently to make any word. Of course, no alphabet is completely phonetic; in the Roman alphabet used in writing English the vowel symbols each represent more than one vowel sound, such as *a* in "father," "sat," and "rate," *e* in "met" and "mete," *i* in "lit" and "light," etc.

Thus the alphabet. Most of this process was performed by the Phoenicians, and as they were the great travelers of ancient days, theirs was the alphabet that had most influence.

Various substances were used to write on. The Ten

Commandments were engraved on two tablets of stone; the laws of Solon, the great Greek law-giver, were carved on planks of wood. Lead and brick also were used. The Chaldeans took bricks or masses of smooth clay and, while they were yet soft, made impressions on them with a metal style shaped at the end like the side of a wedge. The Egyptians made inscriptions on monuments, temples, and tombs. They also used a paper called papyrus which they made from the pith of a reed that grew along the Nile. They cut this into thin strips, laid them side by side, and placing another layer across, pasted them together and subjected them to heavy pressure. Thus a fairly good paper was made, and our very word "paper" is derived from this ancient "papyrus." Before the invention of papyrus the large leaves of certain plants were prepared so that they could be written upon. From this originates our term "leaves" of a book, which in the Latin form, *folium*, has also given us the modern term "folio."

As for the origin of actual paper, there is an old legend that the Chinese first learned to make it by watching a wasp build her nest. The big round nest of the wasp is really made of crisp brown paper. A Chinese watching the wasp one day discovered that she took a bite of wood, chewed it into paste, and smeared it onto her nest. When the paste became dry, it turned into something like our paper. Thus they learned to make paper from wood long before any other people. The clay tablets of Babylon were cumbersome; the wax tablets of Greece were perishable; the papyrus of the Egyptians was fragile; parchment was expensive. It was not until literature was put on a paper basis that democratic education became a possibility.

Word derivations are sometimes quite literal. The roll or *volumen* of papyrus gives us our word "volume" as applied to a book. A book was a roll of written papyrus, and a reader had to hold it with both hands, unrolling it with one hand at one end while rolling it up with the other hand at the other end as he progressed. Old Egyptian scrolls often reached a length of fifteen to twenty feet and were very inconvenient. There were three ways of writing. By the first method the lines followed the paper lengthwise from beginning to end, in which case the reader had to unroll and roll up the paper as many times as there were lines. By the second, to avoid this, the writing was made crosswise. But then one arm had to be held higher up than the other, and that was also tiresome. The third manner was to write in narrow columns more or less of the same size as our printed pages.

A papyrus book was kept packed in a jar. A library consisted of a number of shelves of papyrus jars, each labeled to show the subject treated. There was no paging or index as in a modern book, because the arrangement of written sheets did not admit of this, but there were letters, signs, inscriptions, or labels. Generally the signs or symbols bore some analogy to the subject of the record, and all through the Middle Ages an oak indicated a record relating to forest laws, a head in a cowl referred to a book about a monastery, and the like.

It appears to have been a matter of indifference to Egyptian scribes and readers whether they wrote and read from left to right or from right to left. An ordinary scroll of Egyptian writing shows that the figures of birds, animals, and men all face in one direction, in some cases right, in others left, but throughout any single piece it was consistently one or the other. The writing is always to be read from the direction in which the figures face. The character of the Egyptian writing, consisting mainly of drawing pictures, doubtless made it convenient for the scribe to work in either way. This was not the case with Babylonian and Assyrian writing, which, being made rapidly with the aid of a small stylus, could be more readily carried forward from left to right. Thus the method of writing from left to right gained in general practice.

While this has the approval of all European nations today, they did not adopt this direction of writing at once.

Old Greek and Roman inscriptions have the writing running in opposite directions in alternate lines, leading thus backward and forward across the page in the manner of furrowing a field. Such an arrangement had the advantage that the end of one line so close to the beginning of the next made it easy for the eye to follow on without danger of skipping. In modern books for the blind, in which the characters consist of raised dots variously combined, this method, operated by touch, has been generally followed with satisfaction.

Simultaneously with this system of writing in alternate directions, the Athenian used his writing ability for expressing political preferences. His unusual method was to scratch his vote upon a shell. There was a curious custom established about the time of the Battle of Marathon (490 B.C.) called ostracism, derived from this method of writing. It was a system by which a public man could be banished from the country for ten years. The voting was done on pieces of oyster shell (*ostrakon*), and from this came the term "ostracism." Each voter wrote on the shell the name of the person he would banish and dropped it into an urn. At least six thousand votes were necessary for an election, and if more than half were cast against a man he was obliged to go into exile. The object of this custom was to protect the state against the designs of tyrants by making it possible for the citizens to ostracize anyone suspected of selfish designs on the government. But the privilege was shamefully abused. Some of the noblest and best men that Athens ever produced were driven from their homes, including Themistocles, the builder of the Athenian fleet and winner of Salamis, the most famous naval battle of antiquity, and also Aristides the Just. These things happened for jealousy or other petty and ridiculous reasons.

Even though the Athenian was advanced enough to vote, the idea of separating his words was wholly unthought of at this time. The separation of words from one another, the use of punctuation marks, of capitals, and of paragraphing and similar aids to reading, were all entirely unknown in ancient and early medieval times. The characters went on without any division from the top of the page to the bottom, and the better the manuscript as a work of art, the more uniform and unvarying was the distribution of its characters. There was rarely anything to show the casual observer where one word ended and another began. That this greatly complicated reading can be readily seen by the following example: Thiswillproveaninstanceofwhatismeant.

While the ancient Persians sometimes used a diagonal line, and Phoenician script shows a tendency to mark the divisions between words with dots, the idea did not "catch on." Not until about the tenth century did manuscripts of Europe give evidence of a general adoption of word-spacing, punctuation, capitalization, and paragraphing. In capitalization, indeed, the earlier writing afforded no opportunities, since the Greeks and Romans of classical times and the early Middle Ages used capitals exclusively. Small letters were a space-saving and time-saving invention of the monks in the seventh and eighth centuries. After small script came into vogue, capitals were retained at the beginning of sentences, perhaps quite as much for their ornamental effect as for any other reason. Paragraphing was also a decorative expedient. Word divisions and punctuation were wanted by scribes and readers who dealt with a language not their mother tongue, and these various devices finally took an established place in tradition.

For impermanent notations in ancient times, as when teaching children in schools, sand was scattered on tables, and the finger used to trace forms.

The place of the stenographer and writing machine of today was taken in ancient Rome by slaves, often highly educated, who wrote at dictation. Short notes and letters were written on tablets like our old slates but of wax, on which the letters were traced. For this an ivory or metal stylus was used, with one end pointed like a pencil and the other end broad and flat, like a paper cutter, for smoothing the wax. With this flat end mistakes could be corrected. In fact, the whole letter could be erased and the tablet used again, as was often done when replying to the letter. Our English word "style," as used in rhetoric, comes from the name of the stylus or *stilus* which the Greeks and the Romans used in writing.

Letters of importance were always sealed, as this not only secured them against improper inspection, but also indicated the genuineness of those that were dictated, as autograph signatures had not been thought of. The tablets were put together face to face, with the writing on the inside. Thread was passed around them and through small holes bored through them, and was then securely tied. Upon the knot softened wax was dropped, and to this the seal was applied. When the letter was opened, care was taken not to break the seal. Merely cutting the thread gave access to the contents, and if the letter was preserved, the seal was kept attached to it as proof of its authenticity.

Nineteen hundred years ago the business of publishing a book in Rome was done very much as the gentleman of leisure wrote a letter. Each copy was a separate and independent thing. There were no "carbons" or other duplicating devices of any kind. The hundredth or thousandth copy of the best seller took as much time and work as the first.

When in the first and second centuries there came to be an increasing demand for copies of a work, the booksellers organized methods for quicker multiplication than had been possible when a single scribe was employed to make a copy.

The author's copy would be distributed among a number of scribes, each of whom would write and rewrite the portions assigned to him until the desired number of duplicates had been produced. The sheets were then arranged in order and the rolls mounted in accordance with the methods of "bookbinding" of those days.

Before the binding process, however, the books had to be examined to correct errors. This was a most tedious procedure. Today a proofreader may correct one sheet and know that all duplications of that sheet will be exactly like it. But in those old days each copy had to be corrected separately. No two copies would show precisely the same mistake, as a human agent is not a machine. While nearly all the scribes were slaves, they were well educated. But accuracy was often sacrified to speed, and many deficiencies would be noted.

When this method also proved too slow for the continuing demand, a reader would be employed who dictated at one time to a group of from twelve to fifty scribes, thus saving them the trouble of rolling and unrolling the manuscript rolls, keeping the place, and other incidental time consumers.

The publication of books as a business in Rome began in the time of Cicero. Before that there had been no organized book trade, and no such thing as commercial publication. If a man wanted a book he borrowed a copy from a friend and had his scribe copy it.

There were no copyright laws, and neither the author nor the publisher was protected. The publisher depended, in the case of new books, upon meeting the demand before his rivals, and, in the case of standard books, upon the accuracy, elegance, and cheapness of his copies. He would estimate as closely as possible the demand for any new work that he had in contemplation, would put as large a number of scribes upon it as possible, and would exercise the greatest care that no copy should leave his place until the entire edition was complete. When the copies were once on sale they could be reproduced by anyone, hence the need for this caution. Authors received no "royalties" or specified price for their works, but depended upon gifts or grants from those whose favor they could win by their genius. Going back a little farther, Athenian booksellers derived their receipts not only from the sale or from the hire of manuscripts, but from the reading of these aloud in the shops to hearers who paid for the privilege.

The absence of copyright protection did not work as

much havoc as we might expect, for it was not until the first real paper was made by the Chinese, about A.D. 105, that books could be widely produced. Eighteen hundred years ago, when the Chinese Emperor Ho-Ti desired the favor of Trajan, Emperor of Rome, he sent him a thousand sheets of writing paper, then the rarest gift that even an emperor could offer.

The ink of antiquity was a simple mixture of soot, to which gum was sometimes added. Among Arabians, lamp-black was first made by the burning of oil, tar, or resin, then mixed with gum and honey and pressed into a small cake, to which water could be added when wanted for use. At one time ink and dyes were interchangeable, and the symbology of colors came into play. Only a few colors were recognized, the sense of subtlety in color appreciation coming only gradually to man. According to Bucke, in his *Cosmic Consciousness*, the primitive Aryans, perhaps 15,000 years ago, perceived or were conscious of no color discrimination at all. They saw no difference of tint between blue sky, green leaves, brown earth, and the golden or purple clouds of sunrise and sunset. Philologists find no color word whatever in primitive Indo-European speech. At a later period, but before the oldest literary compositions that have come down to us, the color sense developed into an appreciation of two colors, black and red. Later came yellow, white, then green.

The phrase "purple and fine raiment" that we so often come across had a much broader use than we think, and was not restricted to any one particular color, but denoted the various reds that were then perceived. According to some authorities, it is said that throughout the *Rig-Veda*, the *Zend-Avesta* of Zoroaster, and the Bible the color of the sky is not once mentioned, and was not recognized. The 10,000 lines of the *Rig-Veda*, consisting of Sanskrit hymns and verses, of about 2000 B.C., are largely concerned with a description of the sky; the Bible, it has been calculated, mentions the sky and heaven more than 430 times, but neither of these mentions the color of the sky. Neither do the forty-eight long books of the *Odyssey* and the *Iliad* make reference to it, despite the crystal clarity of blue Mediterranean skies. Four or perhaps three thousand years ago blue was unrecognized, and the subsequent names for it were al' merged in the names for black. Aristotle spoke of the rainbow as three-colored.

This is perhaps a little far afield from the subject of ink, but I think it is an interesting digression as showing how all man's appreciation and sensitivity grows in range and depth and value. It seems to help us realize the immensity of the universe—its latent forces, and the fact that, for all we have and know, the possibilities of what is yet unknown must fill us with eagerness and reverence. . . .

But, getting back to ink, we do not usually appreciate how our world would change without it. Books, newspapers, magazines, would disappear, taking with them the knowledge of the ages and the information of the day. The absence of labels identifying the contents of cans, bottles, and packages would put our supermarkets and other stores into a state of chaos. Paper money could no longer be printed. Today many thousands of specialty inks are used to print on glass bottles, toothpaste tubes, and all sorts of other substances. Byron said something about a small drop of ink that might make thousands think. But that small drop multiplied into countless tons can remake civilization.

The first pen was called *calamus* and was made of a reed brought to a point and cleft much as our quill pens are. Calamus, it is interesting to know, is an accepted symbol of friendship. Walt Whitman, looking for a token of the love of comrades, chose this common plant, a familiar pond reed. "And here what I now draw from the water, wading in the pond-side, . . . this shall henceforth be the token of comrades, this calamus-root shall." Thus as a token of comrades is the symbol taken over in the pen, which has done so much to unite people in understanding and friendship. A word that has come from "calumus" is "calumet," the North American Indian's pipe of peace. Such may be the message of the pen.

Quill pens came into use about the sixth century or a little earlier and held sway as late as 1865. In the leisurely days of the lace ruff and the powdered wig every man was his own penmaker. Cutting quills was an important part of his education. Schoolmasters maintained that much of the charm of the written word depended on the pen that was used, and believed that it should be cut to suit the temperament of the individual writer. Young students were always on the lookout for good goose, swan, eagle, owl, hawk, or turkey feathers. Goose quills made the most satisfactory all-around writing implement. Schoolmasters kept swan feathers for themselves; they fitted best behind the ear, that most ancient of penracks. After a quill was soaked over night, it was heated and scraped to remove the skin coating. Next the nib was cut and split. Then it was reheated to give the desired elasticity and toughness. The knives used in these operations are to this day called penknives. The first metallic pen was made in 1780. It was asserted that more steel is used in the manufacture of pens than in all the swords and guns in the world, thus partly verifying the old saying that the pen is mightier than the sword.

Around the year 1884 it was customary for salesmen to carry on their watch chains a collapsible steel dip pen and a stoppered portable inkwell—an arrangement which kept dry cleaners happy, if no one else (if there were dry cleaners). A few so-called fountain pens had appeared about this time, but they were very crude. L. E. Waterman, an insurance salesman, bought himself one. Shortly afterward, as one of his prospects was ready to sign an application for a large policy, Waterman offered the new fountain pen. But the pen flooded the document and made a mess that angered the customer and lost Waterman the order. Thus it was that Waterman, wounded in pocket and in ego, forsook the insurance business and gave his efforts to the invention of a practical fountain pen. Today ballpoints are cheaper, but they do not respond to a writer as well as the liquid-feed type with its flexible point.

Graphite, from which lead pencils are made, was discovered in Cumberland, England, in 1564. It is not lead at all, and to call it that is a misnomer. It gets its name as indicating the leaden color of writing done with it. Nor indeed is a pencil a pencil. "Pencil" is from the Latin *penicillum*, meaning "a little tail," for the first pencils were fine brushes of hair or bristles, and writing was painted on. The pencil is an invention of very modern times, but it has indeed done much toward the diffusion of the arts and sciences and toward facilitating study and intellectual intercourse in general.

One of the causes of the widespread illiteracy of the early Middle Ages was the scarcity of books. The Romans wrote their books on Egyptian papyrus, but after the Saracens conquered Egypt in the seventh century, papyrus could no longer be had from that country. Parchment was expensive, so books became rare. Parchment was made from the dried skins of sheep and goats, and these skins are thin and pliable. In the ancient world there was a tendency to reserve its use for documents. For many centuries legal documents were always written on parchment, and many are today. A diploma is sometimes spoken of as a "parchment" or a "sheepskin." It was not unusual for a book to be written on parchment from which some former writing had been erased. From this cause many important works of the ancients undoubtedly disappeared.

Even more of the fine literature of ancient days would have been lost to us had it not been for the zealous labors of the placid monks in the dim alcoves of their monasteries, who preserved for us in manuscript form what they had heard of old civilizations or what they had learned by heart or read in forms of writing that have never reached us in the original. All these books had to be written carefully by hand, with pen and ink—a monumental labor. It took so long that one can sometimes trace the passing years in the lives of the scribes. A single manuscript may show the writing growing from strong to shaky as the writer grew feeble; then it changes entirely to that of a novitiate whose chirography is seen to improve with experience. These holy and zealous monks, in their brown robes, bent over a ponderous folio, illustrated their manuscripts with beautiful drawings in colored inks. Such books were called illuminated manuscripts.

Reading and writing at this time were accomplishments wholly confined to the clergy. The word *clericus* or "clerk" became synonymous with penman, the sense in which it is still most generally used. St. Benedict, who lived in the sixth century, did much to favor the preservation of books. Monks were exhorted by the rules of the Benedictine Order to learn writing, and to preserve in the work of copying was considered a service most acceptable to God. Alcuin, the famous monk who lived in the eighth century, said, "It is a most meritorious work, more useful to the health than working in the fields which profits only a man's body, while the labor of a copyist profits his soul." At that time much of the copying was of religious books, in keeping with monastic ideals. Perhaps the claim would not have been made had the books possessed the flavor of many of our modern best sellers.

Ability to write was so unusual that even royalty, presumably with all the available advantages, was without this accomplishment. If kings and great men had occasion to authenticate any document, they subscribed the "sign" of the cross opposite the place where the clerk had written their names. Hence we say to "sign" a deed or a letter. As late as 1516 an order was issued in London to examine all persons who could write, so that the authorship of a seditious document might thereby be discovered. In the eighth, ninth, and tenth centuries books were so scarce and so expensive that none but the rich could afford to own them. A

French countess paid for a single book two hundred sheep and several tons of wheat and other grain.

The art of papermaking was introduced into Spain by the Moors in the eleventh century. From there it spread to Italy and later to France, Germany, and England. Many varieties of modern papers took their names from the early watermarks. The cap and bells, crown, post, horn, and tankard were some of the signs used as watermarks. In 1648 the watermark of the finest English paper bore the royal arms (whence our crown paper), but later in the time of Charles II a fool's cap with bells was substituted for the king's arms, whence our foolscap paper. The original purpose of the marks was to add distinction. Later they were used to prevent forgery of valuable documents or notes. Today they are used largely to designate the manufacturer rather than to distinguish the paper itself.

Like so many one-time, one-use materials, paper has now become indispensable in many uses. True, it has provided the primary medium for cultural advancement and has been the tool of literature, art, and music. In addition, it is used for money, banking and financial transactions of all kinds, insurance policies, blueprints, maps, designs, photographers' papers, government and legal documents, dozens of household papers, waxed paper, packing and wrapping materials and containers, business and social correspondence, and a multitude of products not necessarily concerned with literacy as much as with the daily aspects of living.

Having the ink, pens, and paper, have you ever thought of the story behind the envelope? During the thousands of years man has been sending letters, no one thought of this device in its modern form, although generally there was realized the need for something to keep the contents private and safe in transit. Various expedients were employed to effect this. The ancients who wrote on clay plates sprinkled the plates with smooth powder and then put an outer covering of clay all around the letter. This outer covering was sometimes baked also, or simply allowed to dry, and then the message was ready to be sent. The one for whom it was intended merely cracked this outer clay crust with a tiny hammer, stripped off the broken pieces, and found his baked letter exposed for reading. The powder sprinkled over it had prevented the covering from sticking to it. When papyrus was employed it was covered with strips of the same material placed at right angles to the message and gummed together at the corners.

During the Dark Ages and medieval days, when parchment or a crude form of paper was in use, it was customary to write the letter so as to leave the back of the last sheet blank. This sheet was so folded that its blank side could be used for the address. The corners of the folded message were held together with sealing wax, in those days a very necessary article on every writing table.

Apparently the first man to make the modern envelope was a Frenchman named de Valayer. In 1653 he obtained a permit from the King of France to manufacture and sell in Paris envelopes with stamps on them, or, as we say now, postpaid. The idea, however, was slow in being followed, and evidently it was not until 1700 that England began to use such covers. As late as 1825 Charles Lamb mentioned the envelope as a novelty. And indeed, even at that time, both the French and the English looked upon it as such a luxury that it was made of the daintiest, most expensive paper, and only the wealthiest could afford it. In those old days the flap of the envelope was not gummed, as at present, and every letter writer had to have his sealing wax on hand for fastening even a bought envelope. Only in 1840 was an envelope put upon the market with a piece of gum about half an inch square fastened to the flap. Today we no longer literally "seal" our letters, but the word remains with us from those long centuries when to do so was the custom.

Ever since man began to record his thoughts, whether on planks, stone, papyrus, bark of tree, skin of animal, or tablet of wax or paper, every letter was made by hand. The process was necessarily slow; books were rare and costly, and only the few could have them. But with the advent of printing, a process that would multiply books and make them cheaply rendered learning accessible to the populace. Early printers tried to duplicate the appearance of the written book, even with its abbreviations and illegibility. Aristocrats of the period regarded the printed books precisely as a modern connoisseur of painting regards a chromo —a cheap, meretricious, inartistic imitation, not to be countenanced by a person of culture or taste.

But with the printing press, no longer was it necessary for a man to write laboriously by hand an entire book when he wanted to make an extra copy of it. The setting up of a book in type, thanks to the invention of printing, made it possible to print many copies of a book with the same amount of labor the old monks took to make but one.

The early hand press turned out perhaps fifty or sixty copies of a sheet an hour, but the great presses of today produce a million daily copies of our newspapers. Now the advances in mechanical methods of printing have put low-cost books at the disposal of the poorest of us. Bookmaking is so easy and so rapid a process that publishers are able to keep up with the very latest news and supply further information upon any subject while it is still fresh and interesting. Author and publisher alike are safeguarded by means of copyright, and the author arranges in advance with the publisher a fixed price or rate of royalty for his service. Thus the interests of all are served, and a tremendous power for culture and education is let loose for people today.

So have books, printed matter, and other records accumulated that libraries and archives are becoming unable to cope with them. As usual where a need arises, ingenuity comes along with an answer—in this case, microphotography. First introduced in 1928, this is a method for reproducing documents on 16-millimeter film similar to that used for motion pictures. These compact records on film occupy only 2 per cent of the space required by the original documents, and there are many additional advantages in their use.

Valuable and irreplaceable records of science, history, literature, and art, always vulnerable to the ravages of fire, floods, theft, and other ills, can be quickly, economically, and accurately reproduced on microfilms, which can then be protected in vaults far removed from the original documents. Then, in the event of loss or damage to the originals, one can still have complete and accurate copies produced from the unharmed film duplicates.

But it is not necessary to have a fire or a flood to destroy old records, for the ordinary passage of time causes deterioration. Libraries are filled with newspaper files, rare books, and papers which are crumbling because of age. Microphotography can play an increasingly important rôle in the safeguarding of such cultural records.

The size factor in microfilms alone justifies their use even without all the other advantages. Already large, bulky volumes or magazines can be reduced to tiny rolls of film which can easily be slipped into the pocket and run off at any time on a projection machine. Since they are so small, much material now confined to one place can be sent to scholars and research workers wherever they may be, without requiring them to spend time and money to travel to the original records. Now the original can be kept safe while the duplicate goes traveling. As most research materials come one of a kind, it has never been good practice to lend the originals. Researchers have therefore not only had to travel to the documents, but to obtain manual transcripts of them, which are never fully reliable, or to rely on photostatic or conventional photographic copies, which are reliable but expensive. But through microfilms it is possible for the first time to obtain accurate copies at reasonable cost.

The ever-increasing volume of records used in the conduct of business and industry threatens to bury us under an avalanche of paper. The number of records that must be preserved for their historical, statistical, economic, legal, or other value is so large that the physical facilities available for their preservation are hopelessly inadequate. Microfilming, with its saving of 98 per cent in record-storage space, offers a perfect solution to the problem.

Another saving can be effected in postage, especially if long manuscripts are to be sent over considerable distances by airmail. One company that wanted to send a 2,000-page script saved fifty dollars over and above the cost of the microfilm.

Compact, inexpensive microfilm can give high-school libraries advantages nearly equal to those of present colleges. Colleges and universities in turn can eventually have much of the accumulated wisdom of the world in their microfilm files, notably out-of-print and costly research materials that are now unobtainable.

Truthfully, is it not wonderful that the record of all that interests us of what the world has said and felt and done can be gathered in our own rooms, and that there we may sit and commune with the master spirits of all the centuries? What a splendid point of departure for our own thinking! If each had to begin with his own small experience, how much waste effort there would be! But by means of writing and reading we have the finest basis for building upon the foundations of the past. If we choose our reading wisely, we can go forward from where the past has brought us, and our most valuable asset, the power to think clearly, may benefit from the tonic influence of great thinkers and great achievements of other days.

Epictetus reminds us: "Men are distracted not by the things which happen, but by their opinions about things." The basis for our opinions is thus of great importance. Opinions are one thing we have that we can control. We can think what we like if we are willing to take the consequences, but we can repudiate the thought that is destructive either to ourselves or to others. Reading gives us much of the data for clear thinking, much of the stimulation that of ourselves we would not be able to initiate.

It may be trite to quote Bacon's saying that reading

maketh a full man, writing maketh an exact man. Yet with such fullness and such exactness, we may reveal two essential aspects of life. By reading, we take in, we are receptive; by writing, we are putting out, creating. The impressiveness first, the expressiveness next, interrelated, make a certain balance in us, like positive and negative, action and reaction, the giving and the receiving that go on all through nature. In man's long development from articulateness to literacy he has sensed this feeling, and has employed his ingenuity to further his desire for perfection.

XVIII

When There Were No Post Offices

"NEITHER SNOW, nor rain, nor heat, nor gloom of night stays these couriers from the swift completion of their appointed rounds," reads the inscription over the New York City Central Post Office. It is a quotation from some words of praise given by Herodotus with reference to Persian "postmen" of biblical times. Darius, the ruler of Persia in 500 B.C., devised the way that was followed for a thousand years and more thereafter. He regarded it of utmost importance that orders of the court should be speedily transmitted to the provincial governors, and that their reports should be received without needless delay. Accordingly he established, along the lines of travel already existing between the chief cities of his empire, a number of posthouses placed at regular intervals according to the estimated capacity of a horse to gallop at his best speed without stopping.

At each posthouse were maintained at the cost of the state a number of couriers and several relays of horses. When a despatch was to be forwarded, it was taken to the first posthouse along the route, where a courier received it and, immediately mounting on horseback, galloped with it to the next station. Here it was delivered to a new courier, who mounted a fresh horse and took it the next stage on its journey. Our famous Pony Express in the nineteenth century was little more than this!

In very early times a letter was not necessary in transmitting a message. In illiterate ages message-sticks were used, which, by a code of notches, served merely as reminders to the courier as to what he was charged to say. The first method of spreading news was by word of mouth. This meant that people on foot would obligate themselves to carry news, in the course of their way, though in special cases "runners," whose business was exclusively that of delivering messages rapidly, were employed.

Our word "post" is derived from the upright timber which was a convenient object to which to attach public notices, whence the verb "to post." It came to mean the position or station occupied by the relays of runners.

In Rome, "before the Fall," letterwriting was one of the indoor sports of the man of culture. The absence of telephones and similar devices made letters matters of considerable importance. While there was no public postal service, every Roman of position had among his slaves special messengers whose business it was to deliver letters.

Sending letters by special messengers, especially overseas, was costly, and unless the matter was of great importance the writer was generally willing to wait until another party was going to the desired destination. Any man intending to travel or to send a messenger on a trip considered it a point of honor to notify his friends and acquaintances of this fact in time for letters to be prepared, and people would also transmit letters for complete strangers who might wish for this courtesy.

Naturally there was always the possibility that letters handled in this manner might be lost or delivered to a wrong recipient, and it became the custom to send a copy of an important letter, or at least an abstract of its contents, by another person, and, if possible by another route. The meaning was also frequently purposely confounded by the use of fictitious names known only to the correspondents, or by the use of regular cipher codes. In other words, there was nothing reliable, nothing that one could call a post at all, though efforts were made here and there to set up short mail routes.

Augustus's institution of a state post was intended only for government business. This state post, like so much else in the Roman Empire, was copied from Persia. It was carried on by military couriers. Imperial messengers had their routes mapped out into changes of horses and night-quarters. Posthouses were erected every five or six miles, but only for those who claimed its use by imperial mandate. Pliny, though a favorite and a minister, made an apology for granting post horses to his wife on the most urgent business. Augustus ordered his couriers to ride in company when approaching Rome from different parts of the realm, in the hope that the small talk of strangers might reveal some valuable secrets—detective work of olden times.

The "post" was not a very quick business as we estimate distances today. The messengers covered on foot about twenty-five miles a day. If the distance were great enough, they went by cart at about forty miles. From Rome to Athens a letter took about three weeks on the way.

When the Roman Empire came to its end its postal service went with it, and through all its land each petty ruler had to install his own service. The riders were supplied by the districts through which the road passed. In the ninth century, Charles the Bald undertook to bring order out of the confusion, but the feudal lords were so powerful that he was obliged to sanction their control of the post.

It was only in the thirteenth century that anything resembling a system of public posts was founded in Europe. Then we find the commercial man taking a hand in the postal field, a forecast of the mail as a business getter. The Hanseatic League was an organization of sixty or seventy cities banded together for commercial advantage, and it soon developed a postal department of its own. Since there was little security on the highways, the League took to policing them and gradually became a power that even princes had to respect. The first letter post seems to have been established in the Hanseatic towns in the early thirteenth century as a means of facilitating commercial intercourse. In England a Sir Brian Tuke was mentioned in the reign of Henry VIII as first master of the posts there.

Still another organization instituted a postal service. The great universities entered the field. Students flocked to them from everywhere, and means had to be found for them to keep in touch with their home towns. Else how were tuition fees to be paid?

A new era in postal delivery resulted from the invention of the printing press. The plain man began to read and write. But everything has to be paid for by somebody, and money is always scarce, so the postman's visit was not always welcome, even when he bore no tragic tidings. Sir Rowland Hill, the father of the penny post in England, said: "Postal considerations came to us in a very practical form; every day that brought letters brought also a demand for payment, the postman waiting at the door until he had received his money. In the very early period, when we were most straitened in means, his rap was not always welcome, the demand being certain and sometimes inconvenient, the recompense in the way of news doubtful." And he added a familiar sentiment: "Tradesmen's circulars in particular, which sometimes came from considerable distance, and always unpaid, were great cause of irritation." As the rate varied with the distance, one can understand the hardship that this wrought when the daily amount was added up.

It can hardly be wondered at that there developed evasions and subterfuges, and that letters would be refused on the plea of inability to pay for them, when the handwriting of the address conveyed at least some information, and prearranged peculiarities in it did duty in conveying still others.

In the sixteenth century the time required to go from Strasbourg to Basel, which is only seventy-five miles, and less than the distance between New York and Philadelphia, was eight days by coach. In the seventeenth century a diligence made it in six days. In 1700 this advances to four days and in 1800 by "express wagon" to two days and a half. An Englishman in the eighteenth century could send a letter fifteen miles for a penny, thirty miles for twopence, and so on up, the sum increasing with the distance.

A curious sidelight on postal history might be seen in France during the Franco-Prussian War. The French postal authorities were severely tried at this time, and they resorted to the use of homing pigeons to carry their war instructions to the front. Microscopic despatches were prepared by the aid of reducing devices and could be enlarged at their destinations. It appears that the use of homing pigeons to carry messages is as old as Solomon. The ancient Greeks, to whom the art of training the birds came probably from the Persians, conveyed the names of Olympic victors to their various cities by this method. It also had popularity among stockbrokers and financiers in pre-telegraph days. Early in the nineteenth century the Dutch Government established a civil and military pigeon system in Java and Sumatra, having obtained the birds from Bagdad, and thus revived the training of the birds for military needs in other governments.

Even more colorful is the romance of the post when we trace its development in America. In 1639 the one and only post office in the colonies was the home of Richard Fairbanks in Boston, where were deposited letters from beyond the seas or destined for foreign ports. America with her indifferent roads did little better in colonial times than the old Romans, and in winter it was quite customary for a letter to take four weeks to be delivered from Pennsylvania to Georgia. Correspondents had to depend upon the good will of sea captains, Indians, travelers, special messengers, or the courtesy of Virginia planters in passing the missives from plantation to plantation until their destination was reached.

A monthly post from New York to Boston was established in 1672. And what a hero was our first postman! His road was beset with many dangers: the natural perils of the great, uncivilized wilds, and the treacherous perils which came in the shape of swift-flying arrows from the bows of the red men whose hunting grounds the white-faced intruders were desecrating. Many a story of raging storms and savage beasts, of war whoops and signal fires, could he tell. In the towns and villages through which he passed he became a great favorite. Old and young would gather around him and, while he was distributing his letters, would ply him with questions. What was going on in New York, in Hartford, in Boston? What was the latest news from England and France? He was an honored guest, for in those days news traveled slowly and he was a winged Mercury in comparison. A return letter from New York to Boston within the month! Well, well, well, times were improving!

Benjamin Franklin was the first United States Postmaster-General, the establishment of a general post office having been one of the first acts of the Continental Congress. No one was more interested in the development of the early postal service than Franklin. He had been appointed Deputy Postmaster-General for the crown in 1753, and applied himself so energetically to the work of organization that, despite his various trips abroad, the course of the royal mails continued smoothly. But after twenty years, because of his sympathy with the rebellious colonists, Franklin was dismissed from this office, and confusion in the mails ensued.

The business of William Goddard, a printer and editor, grew so bad as a result that he determined to see what the colonists could do for themselves. He went from town to town in untiring efforts, seeking the establishment of "constitutional" mail routes. Now and then he met with success. Paul Revere was a postrider on such a "constitutional" route and made his famous ride where he was accustomed to carry the mails. Let us turn back to those days one hundred and eighty years ago. As one lolls on the village green outside a gabled tavern, one sees saddle horses and liveried coaches tied to the hitching posts, and hears the men inside making merry over their glasses. Soon a sturdy figure in a tricorn and leathern jacket, high boots and spurs, emerges with a bag on his shoulders. He flings his burden across his horse's neck, springs into the saddle, and is off in a cloud of dust. "There goes the mail," someone remarks.

The busy man of today dashes off a letter in five minutes on his typewriter, signs it with his fountain pen, seals and stamps it, and drops it into the letter chute outside his office door or into the lamppost mailbox on the corner. His great-great-grandfather never dreamt of such a procedure. In his day there was no chute or corner box. He never even imagined a typewriter or a fountain pen. The old gray goose's quills were good enough for him, or rather for the scrivener, for grandfather knew little of the art of the scribe. And the scribe himself, professional though he was, took most of the day to finish the letter. Grandfather sat there the whole time to see that the script was as small as it could possibly be made. Fine flourishes were all right for signing the Declaration of Independence, but not for ordinary correspondence when one had to pay double if one used an extra sheet. The old gentleman wrote seldom and had much to say, so the scrivener was urged to scrimp. When the letter was finished, the sheet was folded and sent by a boy down to the one-room post office. There he left it without paying a cent, for postage was the worry of the man who got the letter, if and when he did; and there it lay until the post happened to leave. The postage stamp appeared only in 1839, the invention of a Britisher. It was introduced here the following year.

Meanwhile, postriders in the more closely settled regions had been replaced by the stagecoach, which could accommodate large stacks of mail. The crowds that gathered about the door of the inn were now multiplied, for when the mail departed it carried passengers. Billowy, bonneted ladies with boxes and bags bustled about, attended by gentlemen in whiskers and high hats. The driver of the stagecoach was a man of large importance in the eyes of those with whom his duties brought him in touch. He enjoyed great consequence and consideration along the road, and had frequent conferences with the village housewives, who looked upon him as a man of trust. In a charming passage Washington Irving says of him, in part:

> The moment he arrives where the horses are to be changed, he throws down the reins with something of an air, and abandons the horses to the care of the ostler; his duty being merely to drive from one stage to another. When off the box, his hands are thrust into the pockets of his great coat, and he rolls about the inn yard with an air of the most absolute lordliness. Here he is generally surrounded by an admiring throng of ostlers, stable boys, shoe-blacks and those nameless hangers-on that infest inns and taverns. These all look up to him as to an oracle; treasure up his cant phrases, echo his opinions about horses and other topics of jockey lore, and above all, endeavor to imitate his air and carriage.

By 1815 there were 50,000 miles of post roads penetrating the wilderness, and some 2,500 post offices. Railroads were at first not taken seriously in connection with the mails. People of 1834 were more skeptical about railway mail service than those of almost a century later were of the newly established airmail. Once a postrider whose route paralleled a railroad track decided to avail himself of the new service and hopped with his bag aboard the train. But his superiors heard of his nerve and with a severe reprimand ordered him straightway back to his horse. He could not afford to run any risk with the mails, they said.

The Post Office Department was getting along well enough in the East when the gold rush to California multiplied its difficulties overnight. Early settlers in California received mail via Panama at a cost of forty cents a letter, but when California was admitted into the Union in 1850 it became necessary to establish an overland route. Ten years later was written one of the most picturesque pages in postal history, when the forerunners of the transcontinental air pilots made possible from New York to San Francisco a fourteen-day delivery, truly winged flight under those conditions. At St. Joseph, Missouri, the mail was taken off the train and thrown across a pony's back. The rider would hop on and away they would go, at top speed for sixty miles, where another horseman would be waiting to snatch the bag and push on. In relays across the plains and mountains the trip was made by these fleet and fearless horsemen. The Pony Express lasted for two years. Not long afterward, the "iron horse" had made its way to the coast.

Times have changed since those early struggles, and today our metropolitan cities are turning out literally millions of letters each day and depending confidently on Uncle Sam to see that they are safely delivered. The post office has added many functions and services to those of past times. House-to-house delivery, special delivery, insured mail, registered mail, parcel post, money orders, postal savings, and always expeditious service are taken for granted. Today one stands at any landing field in late afternoon, and watches airplanes circle around the skies and come to earth to load up again. The atmosphere is alive with craft on wings. Presently several huge, heavy bags are loaded into a plane; a goggled pilot jumps into his seat; the propeller spins; and the plane takes off. "There goes the mail!" someone remarks.

XIX

From Echo to Static

TODAY TELEGRAPH, cable, telephone, and radio have placed the world at our fingertips. No emergency seems so great that we cannot at once communicate a message about it. The wonders of these things still seem like miracles to us, and we look back compassionately, if at all, to those of olden times who lacked knowledge of them.

It is extremely interesting, therefore, to discover that our ancient ancestors did evolve devices that helped them out of many a tight place. Direct verbal or written statement was often impossible. Especially in time of war, information had to be transmitted accurately and quickly to a distance. Letters were too slow; runners not available; other emergencies prevented normal communication. But spurred by necessity, which is ever the impulse to ingenuity, man rose to the occasion and devised such signaling systems as the conditions made possible.

Several methods could be employed, according to the available means for maturing them and with reference to the purposes and needs of the communication. Use was made of sound, as in the case of bells or drums; of sight, as by the swinging of lamps or flags; or indeed by anything that could be perceived at a distance. The chief limitation was that signals had to be agreed on beforehand. While ordinary occurrences could be anticipated in this manner, unusual things could not, so that the most important emergencies were unprovided for. Later, the alphabet could be represented by means of raising torches varying in number of times raised, at either side of a fixed light. Among the ancient Greeks signals were made of a firebrand, waved to and fro to represent hostility or trouble, or, if to signify the approach of a friendly force, held steady.

Nowadays man can speak moderately and a tiny pulsing wire carries his voice across a continent. But once, to make himself heard at a distance, he climbed to high ground and shouted through cupped hands. Mountaintops were of course favorite places for the sending and receiving of news because of their conspicuousness, and when these were not possible, towers were erected.

A great deal of thought was given to the use of fire for signaling purposes, and the Greek historian Polybius made considerable progress with it in the second century B.C. His plan was to arrange two groups of five torches each. To facilitate the understanding of the signals, he divided the letters of the alphabet into five groups. He raised the torches according to a plan that would indicate any letter in the alphabet. If the desired letter was the fourth one of the second group, one torch would show that the second group was meant, and the next would show that the fourth letter was meant. Theoretically this provided for sending any message at all, but it required in practice so many torches and signals that to spell out a few words would take almost a whole night.

The signals of early man were entirely dependent upon clear weather. The signal by smoke or fire could not be sent in the rain; or if it could, it would not be received. Sound signals were more effective as to carrying powers, but even less private. For non-military purposes, in ancient Greece, the town crier was a very splendid person, an officer of the state and municipal government, accompanied always by a musician and using the most flowery, rhetorical language his dramatic mind could conceive. It is only natural that Athens and Rome, possessing two of the greatest and most orderly modes of government the world has seen, should have given to their method of publicity an air of studied dignity and elegance.

Signaling by flags is old, and before this ultra-modern age of wireless, sailors might often be seen to employ this method. The origin of heraldry had its inception in a certain need for signaling. Armorial bearings existed at first in small numbers on the banners of the most important lords, and served as rallying points. As the principle developed, those knights who were not permitted to fly banners adopted colors which helped them to recognize one another at a distance. These colors proved all the more useful when helmets hid their faces. They were worn on surcoats or shields. The more numerous they became, the more necessary it was to vary their field and charges, and that is why they became more and more complicated.

I recently came across a fascinating account of signaling in Cathay as related in 1560 by a Turkish dervish. He describes at length the hardships of the journey, the many months of slow progress, the dangers and adventures. At the border there is a garrison of the king, and the merchants are closely questioned as to what they bring, where they come from, how many of them there are, and other details. The answer being given, the king's guards pass it by signal —by smoke if in daylight, by fire if by night—to the next watchtower, they to the next, and so on, until in a few hours the message reaches the king at Cathay, a thing which would by any other communication require many days. The king sends back his orders in the same manner and with equal rapidity, saying whether all shall be admitted, or only a part, or the whole put off. On their arrival at Cathay a day is appointed for their return, up to which they have liberty to do business. For the people of Cathay do not approve of the prolonged stay of foreigners among them, lest their indigenous manners should be corrupted by some foreign infection.

Signals of light have always been of value for night work, long and short flashes being the chief devices. Color signals may also be devised by means of red, black, and green slides of glass.

It was in 1684 that a Robert Hooke of England invented a system by which messages could be sent a distance of thirty or forty miles. He erected on hilltops a series of tall poles connected by crosspieces above. By means of pulleys he suspended from the crosspieces the letters of the alphabet. In order to read the letters at such distances the eye was assisted by the recently invented telescope. About one hundred years later, what was considered the last word in aërial signaling was the arrangement made by Claude Chappe. On the roof of the palace of the Louvre he erected a post at the top of which was a crossbeam which moved on a pivot about the center. The crossbeam could be moved by means of cords in almost any direction. He invented a number of positions for these arms, each one representing a certain letter of the alphabet. Such machines were put up on towers at places from nine to twelve miles apart, and soon the messages were being relayed over distances of 130 miles. The towers were lighted to be effective at night, and a rapid operator could accomplish about one hundred words an hour.

For close work, when secrecy is wanted, any two persons may arrange a system for the transmission of intelligence between them; secret codes of this kind depend on the inflection of voice, accent on syllables or words, or arrangement of sentences. Sometimes they may be conveniently combined with prearranged gestures. Some have been so elaborated as to create results such as are sometimes attributed to telepathy.

The clear, penetrating tones of bells have been used from the earliest times for signaling people at a little distance, and through the ages have played an important part in many branches of civil, military, and religious life. In remote antiquity hand bells were used in religious ceremonies. In Egypt the feast of Isis was announced in this way. The Greeks employed them in camps and garrisons, and the Romans used them to tell the hour for bathing and for business. The use of bells for assembling political gatherings did not then obtain; public criers and heralds served that purpose.

Bells were introduced into Christian churches in the fifth century in Campania, Italy, from which the name "campanile" is derived, to mean the bell tower. Famous towers sprang up picturesquely on castle and cathedral and "commune." Tall ones, short ones, square ones, round ones, all added their interesting silhouettes to the surroundings. They crowned the center of the building. They flanked the front like sentinels. They ranged themselves in groups at the far end. In Italy they detached themselves proudly from the rest of the building and stood apart, or at most were joined to it only by an arcade.

In medieval times the signal for town gatherings made bell and belfry emblems of communal freedom. Belfries dotted the land until every medieval city of importance that we may visit today proclaimed their usefulness. The most conspicuous were those of the town halls in such cities as Florence and Siena. To mention but a few of the major examples, there is the Campanile of Venice, the Leaning Tower of Pisa, the Bell Tower of Giotto in Florence, the Belfry of Bruges, the towers of Mechlin, Antwerp, Tournai. St. Peter's has its bells; Milan Cathedral has its bells; the Strasbourg spire peals forth. Monastery bells can be heard marking the time on the Lake of Lugano and elsewhere in Switzerland, and church bells resound over some of the long reaches of the Rhine.

By means of established signals, different ideas were conveyed by the bells in these towers. There might be instructions to assemble for a town meeting. There might be an alarm that an enemy was approaching, for the structures were tall enough to serve as watchtowers too, and all had a lookout near the top that afforded a long view over the hills or plains. Another signal indicated simply the hour or the half hour passing into eternity, for the bell was the clock of the Middle Ages. Indeed, our word "clock" comes from *glocio* or *cloche* or *Glocke*, the Italian, French, and German for "bell," as elsewhere noted.

There were many secular uses for church bells, and the chief bell in a cathedral often belonged to the town, not to the cathedral chapter. A storm bell would warn when bad weather threatened. Gate bell would give the signal for opening and closing the city gates. Market bell was the signal for selling to begin. The curfew bell, a corruption of the French expression *couvre feu* (cover fire), was first enforced in England by William the Conqueror as a signal for all lights and fires to be extinguished at 8 P.M. as a caution against fire hazard. In Rothenburg ob der Tauber on a visit I climbed and climbed and climbed to the top of the old Rathaus Turm and found at the top the watchman who still rings the bell every fifteen minutes to advise the town that it is safe!

Again, in olden times, when a person died it was customary to toll the church bell a certain number of times to indicate whether it was a man, a woman, or a child. For a child it was tolled three times, for a woman it was tolled twice three times, and for a man thrice three times. The stroke of a bell when tolling was called a teller, and hence it was nine tellers for a man, or, as folk said in that day, nine tellers mark a man. When this custom went out of use, the quaint expression "nine tellers mark a man" still lingered among the people until its first meaning was forgotten, and it became converted into "nine tailors make a man." Thus it is, through a simple case of mispronunciation, that a tailor has been reduced to one-ninth of a man!

A military use of the bell signal is indicated in the fact that after the taking of Constantinople by the Turks in 1453, the use of bells was forbidden for fear that someone would give the signal for revolt.

The use of a wall for defense had been elemental. The addition of towers to increase its protective quality soon followed. They provided a lookout, afforded a place of refuge, and served other similar useful purposes. The Great Wall of China, to name but one outstanding example, is pierced at short intervals all along its length with towers for just such reasons. Carcassonne was another, and on a smaller scale we find them all through Europe.

The tower, after becoming a material necessity, became a moral need, and offered a certain monumental sumptuousness in religious and also in civil edifices. However its application to the houses of worship of different cults may have originated, whatever may have been the needs to which it at first responded, it was gradually conceded that whether of service or not, it added perceptibly to the prestige and majesty of the sanctuary, and even seemed to symbolize, by its reaching heavenward, a desire toward godliness. Anyway, builders adopted it so generally that it was soon impossible to conceive of a church without a tower.

The most important municipal building, which was called the *hôtel de ville* in France and Belgium, the *Rathaus* in Germany, and the *palazzo pubblico* or *communale* in Italy, was also dominated by its tower. That of the Palazzo Vecchio in Florence is a familiar example, with its square battlements in the building to represent the Guelph faction, and the forked battlements in the tower to represent the Ghibelline, as both parties were too strong for either to be ignored. Very different in type is the great central tower of the Hôtel de Ville in Brussels. Its slender outline and small proportions exclude any idea of defense. The military tower would have a projecting gallery near the top, from the floor of which hot oil might be poured upon an enemy below. But this graceful spire rising evenly upward indicates peace and security. If we compare it with the proud and massive belfry of Bruges, the evolution of towers is quite apparent, for with the change of requirements the character of the tower also undergoes a corresponding change.

Many of our finest buildings have adapted the architectural design or proportions of some of these famous towers as details. A derivation of Giotto's tower is familiar

Official Belgian Tourist Bureau, New York

The Hôtel de Ville and its bell tower on the Grand' Place, Brussels.

to us in The New York Times Building. Madison Square Garden tower, pulled down in 1925, was a replica of the Giralda tower of Seville. The Venice campanile appears in the Metropolitan tower, to name but three. So to bells we owe at least indirectly many of the world's famous towers.

It is interesting to know that bellmakers today exercise reasonable care that bells in one neighborhood shall peal forth in harmonious tones. It can be realized that the discord emanating from various bells as they strike the hour simultaneously would be very unpleasant if this were not done. Thus the bells on the Metropolitan tower at Twenty-third Street, New York, and the Consolidated Gas building tower at Fourteenth Street, can be heard together in certain winds, and are constructed with that end in view.

Bellmaking to a special tone is so difficult that patterns for bell designs are drawn with a knife instead of a pencil to avoid the slightest variation in width of line, the least of which affects the weight and hence the tone of the bell.

The Metropolitan tower bells have a higher mounting than any in the world, hanging at 650 feet above the sidewalk. Voyagers at sea, twenty-eight miles away, and others equally far up the Hudson have reported hearing them. Yet their tones are so soft that they are listened to with pleasure by visitors to the observation gallery one story below the belfry.

Thus throughout the ages bells have served mankind, and the towers, slim and graceful, reaching to the sky and sending forth their message to the people, are reminders in many cities of the twin ideal of use and beauty.

For the more permanent giving of information, rather than the temporary or emergency message, much of real interest may be reported as to the signs of olden times.

Among the charming things to be seen today in old European towns are the wrought-iron craftsman and business signs used instead of billboards for the trades. Silhouetted against the sky, these signs give a gay and gladdening aspect to the streets. In many instances the family's trade is symbolized in a delightful sign of wrought-iron that will hang in front of a shop instead of the modern signboard. The butcher will have a knowing cow or a funny little pig. A pretzel will be the promise of baker's goodies within. I stopped at the Hotel Eisenhut in Rothenburg, and over the entrance was a large iron hat or helmet instead of any printed name. In Germany abstract numbers did not conquer so easily over the more personal and meaningful sign. In the quiet towns of the South, artistic signs keep their place in the sun up to the present day in spite of some ill-advised landlords who think it necessary to hide the humble oxen or lamb in the garret and call their house by some unmeaning foreign name like Hôtel de l'Europe.

It is Falstaff, I think, who says in *The Merry Wives of Windsor*, "Good wine needs no bush." In that age the sign of a wineshop was a bush planted in a tub, and it was meant to convey the idea that if a man sold good wine he need not hang out the bush to declare it. That sign is still often seen, but gayer things are more common. One *Weinstube* in Bavaria has a sign showing a little man drinking, his cup quite upside down as he leans back and seeks to get the very last drop. When competition was trifling and people could not read, some such simple device would do to indicate the trade within.

Many people whose name and business were not adaptable to a simple sign took some arbitrary thing that was pictorial. The animal kingdom was ransacked, from the mighty elephant to the humble bee; the vegetable kingdom, from the palm to the pineapple. Humanity in any quaint form might be seen hanging in the wind. A bank had a sign of three squirrels, whose characteristic interpretation was "May you never want for a nut to crack."

As a neighborhood became more crowded, and there might be two bakers close together, the pretzel would not be sufficient identification for both and they would have to particularize by adding something else, like a pineapple or a black crow.

Elsewhere we see a pig and a dog (not yet hot) running chummily over a sausage store, unmindful of the axes over their heads. German script is often painted on façades and is very decorative, in keeping with certain Gothic features of the town. The inscription here says, "Through the art of the butcher, the pig may be found in the finest company."

Not only stores but houses also needed identification, especially those that served for public purposes. In the Middle Ages the houses of nobility in both town and country, when the family was absent, were used as hostelries for travelers. The family arms always hung in front of the house, and the most conspicuous object on these arms gave a name of the establishment among travelers, who, unacquainted with the mysteries of heraldry, would call them by vernacular names, as Red Lion. Innkeepers then adopted them, hanging out red lions, green dragons, black eagles, or what-not, to inform the public that they offered food and shelter. Sometimes the innkeeper's name might translate conveniently into pictorial representation, as "Breitschwerdt," which means broadsword, but he preferred to be "at the sign of the two lions," assuming the reputation for hospitality of some noble family whose heraldic device this was.

In medieval times these signs not only were charming decorations of the narrow streets but were practical guides for the wayfarer through the crooked lanes. Just imagine going visiting as late as the sixteenth century. You would enter your coach and direct the driver as follows: Eight squares to the Globe Theater; to the right till you reach the sign of the Three Squirrels, straight ahead to the sign of the Red Lion, and then up the hill, turning in four doors beyond the Flying Inn . . . But why go farther? This is enough to show the difference between the past and the present, when we hail a taxi and say crisply to the driver, "Number Ten Downing Street, please."

In our own enlightened times nobody needs the sign as a guide. Everywhere the number has taken its place for this purpose, and we have to admit regretfully, for the sign, the truth of Darwin's words, that while progress in history may mean the advance of thought, it also means the decline of fantasy.

Yet the fantastic is still with us. Electronic devices now available are far more startling than were the telephone and electric light in their day. They will do an incredible number of jobs. They put electrons to work to make radio, sound movies, television, X rays. They also can prevent ship collisions and guide vessels through the fog. More or less they make aviation foolproof and are also used in medical and industrial research. Electronic computers are making massive advances in processing business data.

One of the fascinating servants of man is the photoelectric eye, or phototube. It is a cell so sensitive to light that it will turn on street lights automatically as daylight fades, protect buildings against fires, control street traffic, inspect various products before they leave the factory, rejecting faulty products with an accuracy of measurement of a hundred-thousandth of an inch. It can steer ships, catch burglars, sort fruit, detect smoke, nab speeders, count pills, and match colors with instantaneous swiftness and infallibility. It opens doors, levels elevators automatically—and these are but a few examples of its versatility.

Electronics is truly transforming the world—in industry, in radio and television, in agriculture, in medicine, and hundreds of usages. Under each of these divisions, changes are already in being or contemplated, giving man new insights and powers. All that is needed is the wisdom and good will to bring us to the threshold of a new age of peace and progress. With all the current achievements of electronics already spread before us, we are launched into an electronic age the wonders of which we have hardly begun to glimpse. The electron, a tiny invisible particle of pure electricity—the basis of all familiar matter—contains secrets which we shall probably never exhaust. Man's ingenuity here faces its greatest challenges.

XX

Ingenuity Bursts Forth in Automation

IN THESE DAYS probably nothing has captured our imagination so much as automation. At one and the same time it is seen as the prophet of a new era of boundless prosperity and leisure, and also as the bearer of dismal unemployment and poverty. What is this great new influence which seems to be facing in both directions at once? Automation is the accomplishment of work by a power-driven mechanism entirely without the direct application of human energy, skill, intelligence, or control to the process. It is typified by machines that replace men's muscles, eliminate physical effort, and substitute electricity for manpower. These machines are coupled to devices which eliminate the need for human judgment in the administration or direction of control of the machine—a method known as "machines to run machines." That is, the automatic mechanism can watch what the machine is doing, make sure it follows instructions, and automatically correct mistakes. Another phase of automation is the use of computing and decision-making machines, so-called "electronic brains," which hand out administrative, statistical, and clerical policies.

These amazing devices, however, did not invent themselves, and it is thrilling to consider that human brains are behind these fabulous creations, whatever their repercussions upon our lives.

Many persons have tended to overemphasize the difficulties of adjusting to automation. We cannot attempt to give any exhaustive or complete story of all the ins and outs of the subject. This is to convey the message primarily that if abundance increases through more productivity, it is abundance for all of us. Every American has a stake in increased productivity, because this is the main source of higher living standards. Automation gives us a time of transition from the ancient problem of sharing scarcity to the modern problem of distributing abundance. The new wealth created by the new technology will be very great. In the nation as a whole there will be more than enough new wealth to prevent any group from being victimized while a minority in management profits prodigiously.

Automation is not something that can be performed, developed, or even described in seven easy steps. There is nothing easy about it. The more successful it is to be, the more coöperation it requires on the part of all the people, groups, and factions involved. Executives are fully aware that they would not get from A to B if automation were to take away jobs and incomes, as it is the incomes of job holders which are the main source of the consuming of goods. And without your ability to consume what automation can produce, the whole thing would be self-defeating. Executives are therefore deeply concerned that out-and-out unemployment be kept at the lowest minimum, and that every safeguard be enlisted to avoid it entirely or take it up in the briefest time and with the least possible hardship.

Automation is but one of many forms of technology advancing on a broad front, and there is a tendency to blame it for situations that originate from other causes. It is really but one factor in a most complicated set of relationships that decide our economic well-being. Other factors, such as the effect on our economic security of labor, governmental, and managerial policies and of the international situation, are of equal or greater importance to America's economic health. Automation itself is not one thing, but a highly organized, specialized affair. It depends upon many factors. It should not be regarded as the general menace that has cast fear into many hearts, for it is the source of many and tremendous benefits. Nor is it a universal blessing which we will find waiting for everyone tomorrow morning.

There are many reasons which tend to delay widescale automation, and at worst it is not too great an immediate threat, because of the following conditions. Automation is only applicable in certain special occasions and under a particular collection of situations which occur rather seldom on the whole. The employment of automatic machinery is economically and technically feasible only when a large industrial organization produces a huge volume of standardized goods. This restricts automation at the outset to certain industries which are capable of such massive production and which also enjoy relatively stable levels of sales and prices. Automated equipment must be run at high speed or not at all.

Further, seldom can more than a few phases of production be automated. In one of the automobile companies most noted for it, automation has been applied only in departments which altogether employ fewer than 10 per cent of its workers. Various braking factors slow up the progress of automation. There is the impossibility of meeting immediately the huge demand for electronic control equipment. Another deterrent to rapid change is the enormous cost of installation. Equally crucial is the problem of education. We must greatly increase the current supply of skilled technicians and engineers and mathematicians, particularly specialists in electronics, before even a fair beginning can be made. Automation will therefore develop at a comparatively slow pace, one job, one department at a time. Its techniques are still only partly developed. Automation of even a minor segment of a plant requires much planning, product redesign, consultation with equipment producers, coördination with sales, construction, installation, and many other factors.

It is also true that conventional methods will continue alongside highly automatic processes, just as horses were common on farms a generation after the tractor had come to stay. In some areas, of course, automation may never be economically feasible after its cost is compared with present or alternative means. All these points indicate that it will spread from one field to another over a fairly long period of time, and the industrial and social problems linked up with it can take shape gradually and can be solved more easily. In spite of what some cartoonists may have led us to fear, not a single expert believes that any sudden change is likely to occur in our social or economic life on account of the new techniques.

It is pleasant to realize that both history and experience indicate that automation will help to save and create vastly more jobs, companies, and industries than it will eliminate. Even in the automated plant there will be many workers, technicians, set-up men, programmers, toolmakers, maintenance men, supervisors, and systems engineers. Our industries that have attained the highest degree of automation—automobiles, appliances, rubber, steel, and so on—are also among our largest employers. Let us take one example. Employment has risen in the automobile industry as automation has progressed. In addition, the tripling of the automobile market since 1941 has stimulated employment in many other industries and occupations. The automobile industry is a major user of steel. Capital goods and industrial-equipment manufacturers look to the automobile industry for an important part of their market. Automobile and truck users buy most of the nation's petroleum products. Farmers provide material for plastics and paints. Hundreds of thousands of small business enterprises are dependent on the industry indirectly if not directly, such as those who sell or service autos and trucks, operators of gas filling stations, roadside stands, motels, rubber, etc. Now a gigantic road building program is needed. The waves of job-creating influence flow out from many other industries as well. The manufacture of the equipment itself will provide new employment openings.

As one considers the effect and benefit of automation

on jobs, one should realize that increased productivity is only one of many reasons for installing automatic equipment. Many of these other reasons are such as to improve the employment outlook, and certainly are in no way detrimental to it. They in fact create jobs that were otherwise impossible. Automatic devices may be used because they can act far more quickly than any man could. This makes them necessary as safety devices of all sorts and for control of rapid chain reactions, to protect workers from danger or to warn against it. In atomic plants, automatic equipment can operate where radiation is too dangerous for workers. In many industries automatic equipment handles materials where temperatures are too high for human beings. Where noxious fumes or other hazards exist, machines can do work that men could not. Often the objective is to accomplish something that is beyond the capacity of the human being. With automatic mining machinery and conveyors, iron ore is being brought in under water from the beds of oceans and lakes. There are many other examples of automation doing jobs that would otherwise be impossible, opening up new developments, new industries, new comforts, new efficiency.

Where precise control is important in supervision of industrial processes, automatic controls often can outperform man. In scientific and economic research, electronic computers can make calculations that would otherwise be impracticable. These too increase new job opportunities by creating unforeseen developments of numerous types.

Quality control is the object of many installations. In bakeries, automatic equipment measures ingredients more accurately, and in heat-treating in metalworking industries, it controls temperature and other critical factors, replacing fallible human judgment. In many industries, automatic quality control reduces spoilage and cuts costs.

The decisive fact about automation is that the labor of men becomes many times more productive. Automation will introduce new machines and replace old ones. It may destroy old industries that become uneconomic not just in themselves but in comparison with the better use to which capital and labor can be put. Yet these kinds of changes in the economy are not new. Some activities have been expanding and others declining from the dawn of human invention. In recent years the use of coal for power has been increasingly supplanted by oil, and the next steps may be atomic energy and solar power. Iron was replaced by steel, which in turn is being supplanted by alloys of aluminum and magnesium when lightness as well as strength is important.

Not to "automate" when desirable can create obsolescence, for a business may be left behind if it does not meet competition of those who do automate. Losing one's job for obsolescence is no better than losing it for automation, or a more direct reason. These are familiar labor situations not exclusive to automation. Where automation is necessary or competitively advantageous, it may be the best form of job security. In a country as dedicated to progress as the United States, any enterprise that seeks merely to maintain its *status quo* is almost certain to decline. It will lose its markets to those who do take advantage of new techniques and new methods. Often automation is the means by which a company or an industry is saved, and the jobs that might have been lost are also saved.

It should be noted that had there been no increase in output per worker in recent years, the standard of living would have dropped sharply because there were more consumers in proportion to producers. Without technological equipment the worker today would perhaps be little if any more efficient or productive than any of his predecessors.

Recent years have shown a decline in the proportion of the population available for employment. The birth rate has risen sharply, so that a higher proportion of population is under working age. In addition, youth postpones its entry into the labor force by spending more years in school. And the proportion of population over sixty-five, in retirement or approaching it, is increasing because of increase in life expectancy. People not only live longer but retire from the labor force earlier under prosperity and numerous forms of pensions and retirement income. It also appears that the armed services are likely to continue to be larger than they were before World War II. Accordingly, just to maintain our present rising standard of living and increasingly large proportion of population represented by non-working elderly people and youth, by the defense program and commitments abroad, will represent increased production requirements of some 50 per cent during the next ten years. Only through rather extensive automation can we meet such demands.

Automation and other forms of technology, then, will be needed to raise the standard of living because the labor force at present productivity is not large enough to do all the work required for better living. Facts like these indicate that the 1960's ahead will be a period of both high employment and rapid technological progress. As the country's population will increase much faster than the labor force during the next decade, all available workers, plus advanced technology to increase productivity, will be needed if the standard of living is to continue to improve.

While automation holds out great savings in labor, we need not be afraid of this. We may be thankful for it and devote our energies and American ingenuity to finding the best way of using these savings. They do not need to be taken in the form of unemployment. It is our responsibility to ensure that they are not. They can be taken in higher output, or in more leisure, or both.

Automation has been called a second Industrial Revolution. Today, however, we are not exposed to hardships comparable with those suffered by victims of progress a century ago. Democratic countries have systems of social security undreamed of in the nineteenth century. Even if unemployment benefits do no more than maintain the victims of progress on a bare subsistence level, they save them from the worst hardships to which their forebears were exposed. But modern America is not satisfied with this meager treatment of its workers. The influence of the labor unions will prevent automation from taking place in circumstances that would inflict disadvantages on large numbers of their members. They can and often do slow down the rate of automation and give way only in return for various safeguards. Unions came into existence to prevent the exploitation of man and to improve his way of life. They have built on that original concept until they are a power in the land. They do not waver in their opposition to anything that seeks to deny their members their rightful place in the scheme of things.

Today, because of the existence of a high level of employment, there are incomparably better opportunities for the re-employment of victims of automation. Any unemployment is likely to be purely local and temporary. Pockets of unemployment may exist concurrently with a continued shortage of labor in other districts or industries. When unemployed workers are willing to change their occupations or their locations they can usually step from one job to another. When retraining is necessary, it is becoming more and more customary for this to be financed by the employer, and during the working day, as part of the expense of automation. It does not become a financial hardship for the worker. Many other means are practiced to make necessary transitions smooth and easy as they come up.

Another safeguard is the fact of universal suffrage. Any government that would tolerate a substantial and lasting increase of unemployment would lose votes and power beyond its ability to stay in force.

The advantages of automatic production are by no means of only financial value to the employer. The individual worker also gains. An automatic factory is a much safer and pleasanter place to work in and the jobs are more interesting. There is less risk to life, limb, and health in an automated factory because of distant control of machines and the elimination of the most difficult and dangerous tasks. Fatigue is much reduced, and employees are spared much dull and routine work. The inevitable "pacing" of the operative's work by the conveyor or by the time cycle of the machine can be eliminated when such functions can be made automatic. The uninteresting repetitive tasks will be performed by automation, thus releasing the intricate and wonderful machine that is the human brain for more edifying and elevating occupations. Human capacity is wasted on pushing and lifting, or even on the simpler control functions, if they can be mechanically performed. This costly commodity of human capacity is better employed on work that demands more effective use of intelligence, such as maintenance, planning, and supervision.

Workers have good reason to be thankful to be relieved of some of the tasks they have had to perform before automation. It is said somewhere that automation downgrades the laborer to the status of servant to the machine. The opposite is true. It must cause immense satisfaction to one's self-esteem to possess the power to set in motion and control a vast and highly involved engine capable of the fabulous performances in an automated plant.

Automatic factories will not be workless factories. There will be need for more and more technicians and highly skilled maintenance men. We are already woefully short of such personnel. Statisticians say we need some 25,000 more engineers and 75,000 more technicians each year than we are now training. Russia is training four times this number. Automation is molding the skills of tomorrow. Emphasis is shifting from personal production to personal planning and programming, from workshop to office, from the assembly line to the tool room and draftsman's office. Skills are shifting from manual dexterity and craft skill to skill based on technical knowledge of plant and equipment.

The greatest advantage which man still has over any automatic control yet devised is his astonishing adaptability. Young people can learn a skill that cannot be performed by a machine—for example, be a designer of machines, a repairman, a salesman, or a personnel expert. Or they will go to school long enough to learn to be a manager, an engineer, or a scientist. Or they may get into a trade or profession that cannot be mechanized, such as law, medicine, nursing, teaching, art, or anything that requires imagination, creativeness, or a personal touch.

It should also be mentioned that occupations can and do decline without those engaged in them experiencing any loss of demand for their services. Young workers choose other vocations rather than those that show evidence of decline. With the supply of new recruits cut off, plenty of work usually remains in the occupation throughout the lifetime of those already engaged in it. It is the recruitment rate that declines and if the economy is growing, the transition is accomplished with minimum dislocation. The same holds true for declining industries. Most new enterprises are launched in the growing industries. The number of new newspapers is low, but of television stations high. Every year there are hundreds of new companies organized to make plastics, but very few indeed are formed to operate woolen mills. Job security for those already employed is ordinarily much greater in the old enterprise, even in a declining industry, than in a new enterprise, because casualties of new businesses are highest in the first five or ten years. Here again the effects of automation are lost in the maze of factors which determine business success according to the statistics.

Automation is bound to bring immense material gains to the community as a result of increase in output. This opens a vista of hitherto-undreamed-of expansion in production and consumption. Other thrilling changes are taking place. There is the continuous shortening of the work week, longer vacations, earlier retirement for the old, and longer periods of education for an increasing number of the young. A vast new leisure market will be created, a thirty-billion-dollar market, in such fields as home furnishings, garden equipment, sporting goods, entertainment, travel. Families earn more, and this brings a demand for more and better clothing and higher quality of food. The demand for new homes seems insatiable. Manufacturing industries connected with luxury and leisure are employing more workers.

It has been feared that arts and crafts may get priced out of existence by the reduced cost of manufactured products resulting from automation. To the contrary, a wider demand for craftwork and art is apt to arise as a result of an improved standard of living, which will permit a greater share of consumers to indulge their preference for handmade goods and their desire to possess something unstandardized. On the whole, if only a fraction of what technologists promise for the future comes to pass, within a very few years automation can and should make possible a four-day work week, longer vacation periods, opportunities for earlier retirement, as well as a vast increase in our material standards of living. The next few years will reveal whether the pattern will evolve a shorter workday, shorter work week, occasional longer week end, longer or more-than-once-a-year vacation, maintenance of present working hours with earlier retirement, or various combinations of these possibilities.

We shall be far better able to meet the present and growing social deficits in health, housing, schools, highways, natural resources, and other public services. Through increased productivity, our economy can meet the cost of these long-delayed measures without strain, and their achievement will help, by creating new jobs, to ease any necessary adjustments in employment.

The reduction in the work week will also result in stimulus for additional education, museums, libraries, parks, sports and resort centers. Leisure and only leisure makes possible the development of individuality, spiritual values, intellectual growth, creative talent, and even the cultivation of friendships and more meaningful family unity. Only leisure can liberate the industrial worker from his total preoccupation with toil, the accumulation of money, and the search for security.

There are some who would protest against the inroads of automation. Actually, scientific and technological innovations are neither good nor bad in themselves. Whether these great changes will help or hurt us depends upon how we put them to use. Our goal must be to try to make automation mean what David Morse, Director General of the International Labor Office, calls humanation. "Humanation" means the full release of the creative energies of people. Its emphasis is on the man, not on the machine.

In this connection, government can do a great deal to promote the smooth absorption of automation and other technological changes. Briefly it has three general areas of responsibility: one, to develop and maintain a climate encouraging economic growth and full employment; two, to find out the facts and bring together in a coördinated fashion the picture of change taking place in the economy and to encourage community planning to meet each aspect of it; three, to provide the administrative and social services necessary to minimize economic and social dislocations and individual hardship and to keep these up to date.

Employers themselves are today more conscious of their social responsibility than were the gruff, tough businessmen of the Industrial Revolution. Nor can they ignore altogether the more humane conception that has developed in public opinion. It is realized that "automation is not a device with which to outlaw, displace or dispense with man. It is a means for increasing man's stature and extending his ability to produce in greater volume with less physical effort or mental strain." And the top executives who are thinking along the lines of automation know that they have to proceed fairly slowly and cautiously, not only for their own good, but for yours as well. If automation should merely result in production of more products with fewer persons receiving the purchasing power with which to take the increased supply off the market, then, as the C.I.O. so truly states, it would defeat itself. Automation will only succeed if it continues to increase the demand for the enlarged supply of commodities it makes, and at the same time creates an ever-expanding purchasing power among the masses who make the necessary market.

Executives are therefore fully aware that human planning is as necessary as technical planning. They bear the brunt of planning, decision, and action. Theirs is the heavy load of responsibility for cushioning the impact of technological improvements. Labor unions also have a clear responsibility to promote acceptance of change among the workers. They have to try to combine the interests of their members with the interests of the undertaking and of the community. In these projects there are grounds for cooperation and grounds for conflict. Experience shows that where management and workers have a sense of common purpose and keep each other informed, it is generally possible to work out satisfactory solutions. The companies which have installed automatic machinery and controls most smoothly, with the least difficulty, are those which have had the best labor-management relations. There has been confidence and good will in these relations and a recognition of the problems of both sides. Above all there has been a willingness to look at the problems of the individual workers whose interests are involved. In fact, management, labor unions, voluntary organization, and government services all have something to contribute to the solution of the problems of transition in an automated society.

As Walter Reuther says, we know that the greatest good of society is not served by permitting economic forces to operate blindly regardless of consequences. We now know that economic forces are man-made and subject to controls, that the economic and social consequences of economic decisions can be foreseen, and when consequences threaten to be harmful, preventive action can be taken.

It is, states Mr. Reuther, within our power to see to it that these economic and social changes take place in an orderly and evolutionary manner, toward improved standards of living and social welfare, an extension of leisure, and new horizons of individual opportunities for educational and cultural achievements.

There is every reason to believe that they will so be done.

XXI

Our Debt to Glass

According to an old legend, merchants from a ship of Tyre went on shore to eat. Finding no stones upon which to place their pot, they set it upon blocks of niter taken from the cargo, and used seaweed for fuel. When the fire died down they discovered that the niter had melted and, fusing with the sand and the ashes of the seaweed, had made what we now call glass. Science tells us this method would not have produced heat enough to give the result claimed for it. But whether the story is fact or fancy, it is known that these Phoenicians were noted up to and during Roman times for their glassworking.

Of all the discoveries that came about by accident and were perfected by man's ingenuity, the invention of glass is undoubtedly one of the most valuable. We can hardly realize what our civilization would be without it, and in how many widely different departments of utility, science, and art it is the essential factor. Each of these in turn has had its own results almost beyond our powers to comprehend.

It was the vanity and love of jewels in old Egypt that first created works of glass; its glitter was its obvious appeal. Before one could do anything with it, one could admire it for the strange substance that it is. Early evidence of the domestic use of glass is to be seen in the frescoes of Thebes. On some of the earliest tombs the pictorial adornment includes representations of glass blowers working in quite a modern manner. Remnants of Egyptian glass, beautifully iridescent, have also been excavated from ancient ruins.

This knowledge was passed along in the course of time to Greece, and tombs of Rhodes and Cypress have yielded bottles and cups of glass not unlike those of today. The Greeks in turn introduced cameo decoration to glass. To this class of work belongs the far-famed Portland vase from which, centuries later, Josiah Wedgwood drew the inspiration for his pottery.

Augustus subjugated the Egyptians in 30 B.C. and exacted glass as a partial tribute. Soon it grew so popular that the tribute became inadequate and Egyptian workmen were invited to establish a factory in Rome. Thus was glassworking brought to Italy.

Rome at this time possessed no fine porcelain, and glass was therefore very extensively employed for all sorts of purposes, even in preference to gold and silver. Nero, who was an enthusiastic collector, encouraged extravagance in its use. Before long, mourners were gathering their tears in tear bottles about half an inch high, and placing them beside the ashes of their friends, while Roman ladies and their slaves were carrying their own glass toilet apparatus to the baths.

And where would the modern woman be without mirrors? There is a romance about mirrors that has come down through the ages. Whether or not we all have an instinctive desire to follow the ancient advice, Know Thyself, it is true that men and women of all times have been interested in their reflected images. The myths of Narcissus and of Grecian nymphs who gazed at themselves in woodland pools bear testimony to man's longing to see himself.

Centuries before the present era the Egyptians were using for this purpose disks of highly polished bronze or other metals. It must have been a great moment when milady did not have to go down to the pond to see whether she was looking her best when she had an important engagement on hand. The glass mirror is supposed to be a comparatively modern device, but it is quite possible that such objects were known to the Greeks and Romans, some of whose work was so beautiful, delicate, and modern in appearance. To the Venetian glassworkers of the Renaissance is attributed the revival if not the origin of mirrors, for so many of the ancient arts were lost during the Dark Ages that the same thing may have happened in this particular.

The window pane affords an example of such an occurrence. Think what your home would be without windows, or imagine going window-shopping if there were no glass. Behold our factories today: most of them are real daylight buildings because of window glass.

Some few in ancient times did know that glass could be used in windowpanes, but in that warm climate, there was no point in using it that way, so that all the other things one could do with this fascinating substance became popular, while the idea of window glass held no interest for the people.

The first civilizations developed where it was easy to live in the light—the valleys of the Nile and the Tigris. People could live outdoors. The great exhortation, "Let there be light," was not a serious problem there. Not so in the northern countries, for life was harder in cold climates. It gave the people more zest and more incentive. They came down in hordes to the warmer places, bent on conquest—to gain land in the lighter and easier regions of the Mediterranean.

But these barbarian invaders, with complete lack of foresight, destroyed most of the glassworks, and during the Dark Ages glassmaking became a lost art. At the same time, civilization too became a lost art. Europe for centuries was again the battle ground for people coming south out of the cold in winter, and going north out of the heat in summer.

The homes and castles of those days were not built for the gracious art of living, but on a basis of fear and fortification. Great castles were erected primarily for protection against enemies. They were windowless, except for narrow slits, which were used rather to look for trouble than to admit light or air. Without glass, however, enough air came uninvited, and in the northern climates a window was a drafty and disagreeable hole. Heavy hangings, often of crude animal skins, were used to stuff it.

The early feudal baron of the Dark Ages lived in supreme discomfort. The rain blew in through the window slits; the wind flapped the leather shutters back and filled the long stone halls with icy gusts, blowing what little heat there was up the chimney. As a matter of fact, the window opening was originally located as a skylight, so that smoke could escape from it, and the walls of the building could be better kept solid against enemies. It was only after chimneys came into use that side windows were had at all, and these were placed high in the walls.

The average workingman of today lives in far greater comfort than those medieval lords of Europe, and he has glass to thank for it. If nobles and dukes lived that way in those great, dark, gloomy, man-built caves, the untold misery that must have been the lot of the common people we can only imagine. No wonder these were called the Dark Ages, for dark indeed they were, literally as well as figuratively.

The discomfort of the early window survives in its name, which was originally the equivalent of "wind's eye." In those inclement climates of the North, the people were between the devil and the deep sea when it came to getting light without too much air. And when they made a hole in the wall, the thing that attracted their attention was the fact that what looked in on them was not so much the eye of the day as the eye of the wind. Windows to admit light could only expand when glass became available to keep out the gusts of air. Before glass, oiled paper, skins, horn, and primitive wooden shutters were resorted to.

As late as the sixteenth century, not a single peasant's house in chill Scandinavia had such a thing as a window. In 1521 window panes were so rare that people in possession of them made a point of leaving them to favored heirs in their wills. At that time, a wooden shutter to be closed over the window opening at night was considered so necessary to comfort that a traveler carried one along in his luggage.

In many modern places throughout Europe, one still finds this use of shutters instead of glass. A few steps from the Ponte Vecchio in Florence, each noon sees the complete shutting-up of shops for the midday siesta. The shop's salespeople do not go for luncheon in shifts, as we do here, but close up the place as tightly as we would for a long vacation. It makes a gloomy effect, and I remember how aggrieved I felt that I could not even window-shop from the outside! Another arrangement that I saw instead of glass windows was in the little town of Ponte Delgado in the Azores. Here wooden screens were used at the shop windows. During the day they were lowered to form a counter running across the store, the customer standing outside. That system would hardly do for some of our modern emporiums. Nowadays the whole nature of advertising would have to be altered as far as shop-window displays are involved, if we had not glass to reveal, yet protect the contents.

When glass windows became general, man did not have to go south in winter to keep warm, nor north in summer to keep cool. He could build in one place, become a home-owner, and create a civilization. With the coming of glass, man could live in the light and keep out of the cold at the same time. It was a tremendous advantage. Glass windows became the important feature in architecture. The temples which had been windowless and lighted with the feeble glimmer of candles, or had worried along with uncovered skylight holes that let in all the rain from heaven, now began to have beautiful domes designed for them with vertical windows of great size to let in the light and to keep out the rain and cold.

Glass was not used for domestic windows in England and France until the twelfth and thirteenth centuries. At this time it was still so difficult to acquire, and so expensive, that only the wealthy could have it, and even they took down the glass and packed it away for safety when they were to be absent from home. No handsomer present could be made to a bride at her wedding than a whole sash of glass which would enable her to have daylight in her bedroom in all kinds of weather. In those pre-electric days, artificial light was so poor that it only "made the darkness visible," and the light-bringing value of glass was therefore especially important.

Toward the end of the sixteenth century, glass grew cheaper and was more generally used throughout the Continent, but in the early part of the century it was still so difficult to get that a Danish king, in his plans for a new palace, specified that but one part of it be furnished with windows of glass. Their cost at that time caused them to be firmly fixed into the wall to avoid the risks of opening and shutting. More light therefore meant less fresh air.

Window glass gradually became common enough by the eighteenth century so that even people in moderate circumstances could have it. But the panes were small and so uneven that objects seen through them were fantastically distorted. It was only in the late nineteenth century, when plate glass was manufactured in large sheets, that there was anything approaching modern comfort at a cost within reach of those who wanted it.

Many were the changes effected by the revival of window glass. Architecture itself, and our resulting ways of living, have undergone many improvements as a result. Industrial changes have also occurred. Before window glass became a common possession, the air which came through the open windows was beneficial, but the cold or wind were not always so. If the outdoor temperature was not too low and the wind not too brisk, life and industries within could go merrily and well, but let either the air or its temperature be unfavorable, and at once discomfort followed. Think of the demoralizing effect upon a factory, or even a home, in the face of such uncertainty. Imagine gusts of wind lifting up your materials, or rain pouring in to complicate matters. Modern inventions could hardly be effective to the same degree they are, without these hardships cared for.

The great change from home life to industrial life that began in the eighteenth century could not have gathered its present momentum until an abundance of inexpensive glass filled homes with a flood of daylight, as well as shops, offices, and factories, at the same time keeping in the artificial heat. Beginning then, outdoor life rapidly lost its devotees, while the world of indoors gained people, willingly or unwillingly, by thousands of thousands. A host of new industries sprang into being in the wake of window glass, and these in turn resulted in other industries, scientific inventions, and discoveries. Large factories were made possible; big business began; and the physical conditions of home life were altered.

In the beginning, some of the advantages were counterbalanced because windows were fixed rather than movable. The air which all breathed in home, shop, and office became less pure and was often flecked with the dust of industry. Fixed window panes seriously hampered ventilation, and people continued afraid of "poisonous night air." They became softer, having less need to resist the rigors of outdoors. When one considers basic changes in light, temperature, humidity, composition of air breathed, and readjustment of the person to disease germs—well, it is clear that glass did quite a little changing of our environment. Today specially treated glass admits special health-giving rays of light, and ventilation engineering, air-conditioning, and the like have been perfected to overcome such ills.

Windows of a house of the present time, made of glass and putty, of weather stripping, of patent locks, of sash weights, or steel casements, are no longer the eyes of the wind. Rather, they have become the eyes of the house, one of the expressive features that give it character and reveal what it is like, even as the eyes of a person are "windows of the soul."

Not the least important result of window glass is that visual and mental horizons are extended, and habits of thought correspondingly altered, so that even the broader aspects of civilization have resulted from this marvelous substance. Well may we speak of magic casements.

Without windows, a house is a prison-like affair, lifeless as a tomb, shut in and limited. It might be lighted artificially, and ventilated by the latest mechanical means and air-conditioning, but it would lack any spiritual link between the indoor life of the man and the outer world which real windows provide. Windows, then, offer a profound spiritual quality, giving us a sense of unity with the rest of the world, together with physical comfort and a sense of mental security and peace.

Outside the home too, window glass has its uses. Our much-vaunted speed would be greatly hampered if the men who run our twentieth-century fliers could not look ahead through a transparent window pane, or protect their eyes with goggles. And if the windshields on our motor cars were translucent or opaque, how fast could we go if we had to look over the side in order to see clearly? At sea, the accuracy of our compasses and other instruments would be reduced if they could not be protected with glass. Without glass our towns and cities would be places of gloom, for not only window glass by day but electric light by night depends vitally upon this substance. If we did not have the signal lights announcing danger or safety, how we should have to grope our way instead of moving quickly and confidently along!

Nowadays we have special types of glass for safety in such places where flying or breaking glass would be dangerous. For airplanes, trains, motorcars, and the like, for goggles, for bullet-proof bank inclosures, a non-shatterable glass is both a great asset to safety and an economy as well. Such glass may crack if subjected to a powerful blow, but the danger of flying fragments is eliminated.

No less important are the revolutionary effects of glass when employed in the use of spectacles. To be without them would undoubtedly be a hardship for this print-devouring world. A large proportion of people would be unable to read ordinary books, newspapers, and correspondence. Others might be condemned to constant headaches, while many children having congenital optical defects might never learn to read and write.

The use of spectacles in Europe dates from about 1300, but at a time when the merest minority of people could either read or write they were considered vain and luxurious novelties. A certain snobbishness was ascribed to those who affected them, as if they were showing off their learning.

Early eyeglasses were made from a kind of smoky crystal, berillus, whence the German name *Brillen*. They are thought to have been invented by Salvino de Armato of Florence toward the end of the thirteenth century. In a recent performance of the opera *Aïda*, the setting of which takes place in ancient Egypt, a near-riot was caused behind the scenes when one of the Ethiopian slaves started to march on to the stage with eyeglasses perched upon his nose. It would seem an insignificant thing, but not when one considers how glass catches the light and reflects it. The anachronism would have been noticed by observant watchers in the topmost galleries.

I am saying nothing about the decorative beauties of glass—the wonderful bowls and vases, the table appointments, and the rare craftsmanship produced by the Venetians, Bohemians, the French, the Swedish, and others. It is interesting, however, to tell the reason why the name *fiasco*, which is Italian for "bottle," is identified with our idea of a failure when a plan or a performance turns out badly. The Venetian glassblowers, in making their exquisite glassware, when they detected a flaw in the bulb, could not

Trans World Airlines, Inc.

The heavy façade of a medieval palace at Milan, Italy, broken only by small, narrow windows, as was common before the general use of window glass. (The fountain is modern.)

use it for the original purpose; but not wanting to waste it, they converted it into an ordinary bottle, or *fiasco*. Thus the word "bottle" became synonymous with failure.

Glass was always loved for its unusual sparkle. When formed after the only known models for drinking vessels, such as gourd, coconut, or horn, a common feature of these Saxon devices was that they could not be set down until emptied. They were literally "tumblers," and when the time came that they did not tumble, the name survived. A tumbler that would stand alone made an agreeable change in table manners and in comfort. When there was a foot to hold it upright, it was no longer imperative to drain the contents of the glass in one gulp. To this day an upstanding glass is called a goblet, which means a little gulp. The use of glass for table utensils was common enough in the thirteenth century to allow drinking vases to be known under the name of glasses, no matter what they were made of. In the reign of Louis XI, in order to hide silver plate, which was being commandeered by the king, ewers and goblets of glass tended to become still more general.

In modern life, glass is not so essential for such utilities as it is in serving the purposes of science. A world without cameras, photographs, moving pictures, or television would be a very different place from the one we know today. Much would be lost to us of pleasure and information, and that these are instruments of utility for general record-making is of course obvious. But without glass they would never have been known. It was the glass prism that enabled Newton to discover the composition of light and develop the science of optics.

When the Space Age looks for a material that will not burn, corrode, absorb moisture, sag, or lose resiliency, that insulates against heat and sound but not radar, is light but won't shift out of place, won't expand, it thinks of glass. Obviously much research has been and is being done to produce various types of glass with these many qualities.

There is evidence that the Chaldeans were acquainted

with the use of magnifying glasses some 6,000 years ago. Undoubtedly the focal properties of convex "burning glasses" were known in very remote times, but in the ancient world generally a glass lens appears to have been treated more as a philosophic toy for producing striking and often curious effects rather than as an aid to vision.

Most wonderful of all, perhaps, is the way in which glass, in the form of lenses, has enabled man to solve some of the riddles of the universe. The marvelous beauties all about us that are unseen by the unaided eye are revealed in all their mystery by means of the microscope. The little drop of water reveals itself a miniature world, peopled with millions of tiny organisms, living their own lives and carrying on the great laws that move harmoniously through all nature. Those other worlds whose minute inhabitants so influence our destinies can lie before us, can enlist our intelligence to increase their usefulness, can thrill us as they show the underlying unity of cosmic law. These little universes, patterned exactly on the laws of the great universe of the stars, the relationship between the microcosm and the macrocosm, give a lesson so glorious that we can barely comprehend it.

The microscope is used in countless ways. It tells us secrets of plant and animal life; and it may assure us pure foods, unadulterated textiles, honest tools. It shows us the structure of rocks, from which we read the history of ages. It aids widely in medical research to safeguard life and health. It teaches us that of all the greatly maligned bacteria of which so much is heard, for every one of the known harmful ones there are millions that are not only harmless but useful, and many give enchanting beauty to these invisible worlds made visible. In agriculture it has many uses, and helps us in our understanding of the way things grow. For chemical progress the microscope is indispensable. In physics it has taught us the structure of atoms and their nature as small universes. In manufacturing, it is a constant safeguard. Electronic microscopes can make enlargements of

200,000 diameters, revealing new secrets as to the nature of matter. And these are but a few of its marvels impossible to know without glass.

The same lenses, combined somewhat differently in the form of the telescope, reveal the mysteries of the firmament. Now we can detect a sunspot and predict with some certainty a famine in India from its effects, or foretell the coming of an eclipse. And how wonderful is this alone, as showing us that the laws of the universe are reliable and dependable, inexorable, and that we have only to work with them and not contrary to their nature! The exact balance of centrifugal and centripetal action, as we see it there, is an example of the perfection of the law, and gives us hope that we may some time learn to regulate our own small universes by keeping similar perfect poise and balance, come what may.

The two studies together, of the infinitely great and the infinitely small, give proof of the ancient teachings, "As above, so below; as within, so without." This little sentence contains a wealth of wisdom and philosophy. In your excursions into the practice of ingenuity, keep it in mind.

Before the invention of the telescope there was certainly a science of astronomy, but so inadequate, so much a matter of unprovable guesswork, that it hardly merited the name. The greater part of our knowledge of the science is due to the telescope, which, by magnifying our image of the celestial bodies, enables man to study them at much less than their actual distance. The first instrument was constructed in 1609 at Middleburg, Holland, by an optician named Lippershey. The news of its construction reached Italy in the summer of 1609, when the idea of constructing one of his own presented itself to the fertile mind of Galileo. The subject fascinated him, and he made several, each improving on the last, and giving longer and longer ranges. With the third, he commenced observations of the moon. With his fifth he discovered the satellites of Jupiter. The story of the hardships he endured as magician and as disprover of Aristotle is well known.

Ever since, some have lamented the passing of the old idea of a little earth with man its king, and a history of a mere few thousand years. Others have felt their religious faith shaken by the marvelous revelations of science. But there is no conflict. While modern astronomy has revealed a universe infinitely vaster than our forefathers dreamt of, it has correspondingly exalted our knowledge and faith in the Creator of it. And is there not also great comfort in the thought that the human mind, so infinitely small and insignificant in space or time, is yet able to span the vast worlds of the universe, to weigh the stars, to predict the movements of the heavenly bodies? If man's ingenuity has done this, where shall it stop? Truly it gives a sense of dignity to the human intellect that can soar above its environment into these regions of perfect harmony and order where, undisturbed by earthly conflicts, all expresses the eternal laws of goodness and beauty.

Index

www.ingramcontent.com/pod-product-compliance
Ingram Content Group UK Ltd.
Pitfield, Milton Keynes, MK11 3LW, UK
UKHW020204290926
15470UKWH00001B/14

9 780837 180410